D1488256

Principles of Pleistocene Stratigraphy
Applied to the Gulf of Mexico

Principles of Pleistocene Stratigraphy Applied to the Gulf of Mexico

edited by
Nancy Healy-Williams

International Human Resources Development Corporation   Boston

Cover design by Diane Sawyer

Interior design by Susan Swanson

Library of Congress Cataloging in Publication Data
Main entry under title

Principles of Pleistocene stratigraphy applied to the
    Gulf of Mexico.

    Includes bibliographical references and index.
    1. Geology, Stratigraphic—Pleistocene.   2. Geology—
Mexico, Gulf of.   3. Petroleum—Geology—Mexico, Gulf of.
I. Healy-Williams, Nancy, 1950–
QE697.P84   1984   551.7'92'0916364   84-6669
ISBN 0–934634–72–6

Printed in the United States of America

Geological Sciences Series

Contents

Preface

This book evolved from a workshop entitled "Recent Advances in Pleistocene Stratigraphy and Paleoenvironmental Interpretation" held September 19–22, 1982 at Hobcaw Barony in Georgetown, South Carolina. Hobcaw Barony is the site of the Belle W. Baruch Institute for Marine Biology and Coastal Research of the University of South Carolina. The workshop was convened to familiarize petroleum geologists and biostratigraphers with advances made in the past decade in high-resolution stratigraphy of Pleistocene marine sediments. For that workshop, short-course notes and an extensive bibliography were provided to the participants. This book is a direct result of those short-course notes, with a significant expansion of all chapters, including an updating of references and illustrations.

Topics covered within this volume are magnetostratigraphy, planktonic foraminiferal biostratigraphy, oxygen isotope stratigraphy, tephrochronology and a review and updating of terrestrial-marine correlations during the Pleistocene. Versions of the chapters on oxygen isotope stratigraphy and tephrochronology will be appearing in the forthcoming third edition of *Subsurface Geology* by LeRoy and LeRoy and appear by permission of the Colorado Schools of Mines.

This work would not have been possible without the aid of our co-workers, both at the workshop and in our laboratory. In particular, we would like to thank David Mucciarone, Mary Evans and Karen Gruenhagen for their technical support. Special thanks go to Donna Black for her diligent typing of the manuscripts. Helpful discussions with various workshop participants are also acknowledged, particularly those with Barry Kohl of Chevron, Inc. We would also like to thank George deVries Klein for his encouragement in putting together this volume for publication. Much of the published and unpublished data presented in this volume have been the results of National Science Foundation–supported research.

Nancy Healy-Williams

Introduction

The Gulf of Mexico is one of the world's major oil provinces. Mesozoic and Tertiary deposits are found in the Gulf of Mexico with thicknesses ranging up to 22,000 meters (about 66,000 ft.). As exploration moves progressively offshore into deeper waters, thick sequences of younger sediments (upper Tertiary to Holocene) are encountered that contain major petroleum reservoirs. As shown by Holland et al. (1980) the reservoir of one of the largest producing oil fields on the United States continental shelf (Eugene Island Block 330) contains oil that must have migrated during the last 500,000 years. Foote et al. (1983, p. 1053) also reported that "traps formed even as late as middle Pleistocene would have been ready to accumulate migrating oil and gas in the Holocene." The Pleistocene, however, represents a time period that is often difficult to correlate precisely from section to section owing to the lack of adequate biostratigraphic and lithostratigraphic markers. The shortcomings of the time control are primarily due to the fact that the Pleistocene represents a relatively brief period of geologic time during which a huge accumulation of terrigenous clastic sediments occurred.

Because of the discovery of, and the potential for, commercial quantities of petroleum in Pleistocene continental margin sediments, we feel a strong need exists for a book discussing new techniques used in determining high-resolution stratigraphy in Pleistocene sediments. Over the last two decades, many researchers have been utilizing deep-sea core material to gain an understanding of the history of the Pleistocene. To perform this work, techniques had to be developed for time-stratigraphic resolution better than ±50,000 years. Many of these techniques are very new and have not received wide distribution to researchers in petroleum exploration. The goal of this book is to introduce these recently developed topics and techniques to the petroleum geologist.

The volume does not purport to outline the geological and geophysical aspects of the Gulf of Mexico, nor does it cover such related topics as seismic stratigraphy, lithostratigraphy, or well logging. Rather, the book is an explanation of the stratigraphic approaches that are of potential use in exploring for petroleum reservoirs in the Gulf of Mexico. For further reading, see the classic papers by Uchupi and Emery (1968) and Ewing et al. (1962). For a review of the physical oceanography of the Gulf of

Mexico, see Capurro and Reid (1972). Topics dealing with the geology and geophysics of the Gulf region are summarized in Resak and Henry (1972), Bouma et al. (1978), Powers (1981), and the references contained therein.

We hope that this book will serve as a resource and reference for petroleum biostratigraphers, sedimentologists, and others involved in the many aspects of exploration and production. We also feel that the book will be of use to students of marine sediments, marine geology, and continental-margin stratigraphy, in general.

The book is divided into five related chapters, each of which describes the techniques involved in stratigraphy of Pleistocene sediments. The emphasis is on the Gulf of Mexico, but the techniques described can be applied to other marine sedimentary basins.

In chapter 1, on magnetostratigraphy of marine sediments, Michael Ledbetter demonstrates how magnetostratigraphy can be used to provide first-order dating of marine sequences. Radiometric dating of terrestrial lava sequences establishes the timing of reversals of the earth's geomagnetic field. The reversals may be recognized in marine sequences by examining the polarity of the inclination in unoriented drilled cores with top and bottom integrity. Magnetic polarity boundaries provide the framework for estimating sediment accumulation rates, recognizing repeated sections, dating seismic reflectors, and locating unconformities. Magnetostratigraphy furnishes a convenient method of determining the absolute ages of second-order stratigraphic horizons such as biostratigraphy, oxygen isotope stratigraphy and tephrochronology. This chapter establishes the time scale used in succeeding chapters.

Chapter 2, by Robert Thunell, deals with high-resolution biostratigraphic correlation of Pleistocene marine sequences, in particular those of the Gulf of Mexico. Outlined in this section are the major steps involved in establishing and interpreting Pleistocene planktonic foraminiferal records. A quantitative micropaleontologic approach provides the potential of establishing a high-resolution biostratigraphic scheme for the entire Pleistocene of the Gulf of Mexico. New data are presented that also illustrate the applicability of quantitative faunal analyses in improving paleoenvironmental interpretations of Pleistocene Gulf-Coast sequences.

Chapter 3, by Douglas Williams, builds on the stratigraphic control outlined in chapters 1 and 2 and demonstrates how oxygen isotope stratigraphy can be used to tie continental-shelf and -slope sedimentary sequences to an absolute time-scale for the Pleistocene. The oxygen isotope record of the Pleistocene is one of the tools most widely used by researchers today for making global stratigraphic correlations in marine sediments. Oxygen isotope stratigraphy also can place depositional cycles in the framework of glacio-eustatic sea-level changes in the Pleistocene and test the synchronous nature of local biostratigraphic and lithostratigraphic horizons in the Gulf of Mexico. The chapter by Douglas Williams includes newly obtained isotope data from sequences of the Gulf of Mexico.

Volcanic ash layers provide an excellent time-stratigraphic horizon in the geologic record, owing to the rapid accumulation rates of single eruptive units. In chapter 4, Michael Ledbetter describes how ash units can be utilized to develop a regional tephrochronology framework. Recent work has shown that the electron microprobe can be used to obtain a geochemical "fingerprint" of glass shards from well-dated sedimentary sections. Ash layers in undated sections can then be correlated with these geochemical fingerprints. This technique allows the stratigrapher to apply an established tephrochronology to sections with dispersed tephra far from a volcanic source. Within the Gulf of Mexico, tephra layers provide a valuable means for the regional correlation of late Pleistocene marine sequences. Combining tephrochronology with oxygen isotope stratigraphy and biostratigraphy provides the potential for a well-controlled stratigraphy of marine sequences.

In the final chapter, Richard Fillon discusses recent advances in Pleistocene terrestrial-marine correlations. Continental-margin sedimentation in areas of large terrigenous sediment input responded strongly to the dramatic environmental perturbations of the Pleistocene. Thick deposits of sediments with great resource potential accumulated in numerous depocenters on margins off major river systems during the Pleistocene. In this chapter, Fillon discusses the problems inherent in using linear interpolation for correlation purposes in complex continental-margin depositional environments, especially as related to the Pleistocene sections of the Gulf of Mexico. He also addresses how the classical midcontinental glacial-interglacial stages relate to the latest theories of global climatic change and how these recent studies provide a glacial climatic model for the last 5 million years.

The emphasis of this volume is on the Gulf of Mexico, but the stratigraphic approaches discussed are not confined to the Gulf of Mexico in their scope or applicability. These techniques can be applied to most other basins containing thick sedimentary sections of Pleistocene age or older. We hope that this volume will provide a framework for analyzing deep-water marine and continental-margin sequences from the Gulf of Mexico and other sedimentary basins by means of high-resolution stratigraphy.

Principles of Pleistocene Stratigraphy
Applied to the Gulf of Mexico

1

Pleistocene Magnetostratigraphy

Michael T. Ledbetter
Moss Landing Marine Laboratory
San Jose State University
Moss Landing, California   95039

1 .

Introduction

Numerous studies have documented that detrital magnetic minerals (iron titanium oxides) in sediments record the orientation of the Earth's magnetic field at the time of deposition (Harrison and Funnell 1964; Harrison 1966; Opdyke et al. 1966; Hays and Opdyke 1967; Watkins 1968; among others). Because the Earth's magnetic field has reversed its polarity (direction) many times in the geologic past, the pattern of normally and reversely magnetized intervals in a sedimentary section provides a globally synchronous stratigraphic tool for dating sedimentary sequences, estimating sediment accumulation rates, recognizing repeated sections, locating unconformities, and dating seismic reflectors. The principal utility of the paleomagnetic method to the Pleistocene stratigrapher is to determine the age of biostratigraphic and lithologic horizons within a sedimentary sequence or to place another type of stratigraphy, like oxygen isotope stratigraphy, into a chronostratigraphic framework. Calibration to an absolute time scale is possible because the geomagnetic field completely reverses its polarity within a period of 1,000 years and has undergone four polarity changes in the Pleistocene (0.72, 0.91, 0.99, and 1.66 million years before the present [MYBP]) and numerous times in the preceding few million years (fig. 1.1).

The utilization of magnetostratigraphy to assign ages to Pleistocene (and older) sedimentary sections is now a routine procedure for marine and terrestrial stratigraphers (for example, Lowrie and Alvarez 1981). The paleomagnetic time scale was first developed by compiling the magnetic polarity *and* radiometric age of samples from widely separated terrestrial lava sequences (see Watkins 1972 and Opdyke 1972 for the historical development of the time scale). Lavas were used to establish the time scale, because $^{40}K-^{40}Ar$ radiometric dating techniques could be applied to the same samples used for magnetic polarity. The amalgamation of many dated sections worldwide was necessary, to acquire a complete Pleistocene record. This is due to hiatuses in lava extrusion at any one site. The paleomagnetic polarity time scale is well established now for the Pleistocene, and only short-duration ($< 10,000$ year) events are left to be delineated. The time scale has not changed significantly since the one compiled by Cox

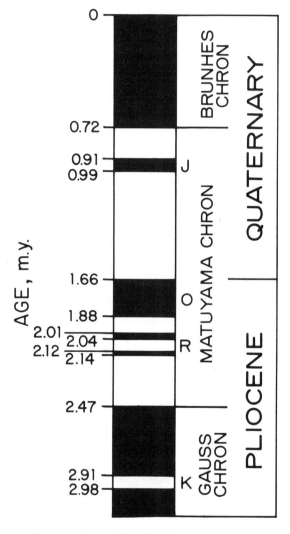

Figure 1.1 *The magnetostratigraphic time scale of polarity reversals in the Earth's magnetic field. Four reversals have occurred in the Pleistocene Epoch (Mankinen and Dalrymple 1979). By convention, times of normal magnetic polarity are represented by shading, and reversed polarity is shown by white intervals.*

SUBCHRONS: J = Jaramillo
O = Olduvai
R = Reunion
K = Kaena

(1969), except that the decay constants used for the K–Ar series have been revised slightly (Mankinen and Dalrymple 1979).

The latest version of the paleomagnetic time scale for the Plio-Pleistocene (fig. 1.1) shows that the last 3 million year record is characterized by (1) a long period of normal polarity with no major field reversals (called the Brunhes Epoch or Chron), (2) the Matuyama Reversed Chron with three short-duration subchrons of normal polarity (Jaramillo, Olduvai, and Reunion), and (3) the upper portion of the next oldest normal chron (Gauss), with one reversed subchron called the Kaena (fig. 1.1).

Once the polarity history of the Plio-Pleistocene was determined, it was then possible to use the magnetic anomaly patterns surrounding the mid-ocean ridge spreading systems to derive an independent time scale (Vine 1966; Heirtzler et al. 1968). Assuming a constant spreading rate, the ages of the magnetic anomalies are assigned by correlation to the paleomagnetic time scale for the period 0–5 MYBP and extrapolated beyond 5 MYBP by using the assumed constant spreading rate (fig. 1.2) (LaBrecque et al. 1977; Ness et al. 1980). For the Pleistocene portion of the time scale, the ages of magnetic reversals are nearly identical in both time scales because the assumption of a constant spreading rate is valid. As age increases, however, the anomaly time scale may be in error, owing to the extrapolation of spreading rates beyond the oldest anomaly correlated to a radiometric age, and the user should be aware of the inherent assumption of constant spreading rate in older portions of the anomaly time scale.

In areas of constant (or nearly so) sedimentation rate, the magnetic polarity pattern in the sedimentary section under study may be compared directly to the time scale, and a correlation determined. This correlation is made possible by the unique pattern of reversals in the post-Pliocene record (fig. 1.1). In areas of widely fluctuating sedimentation rates and/or hiatuses, however, the correlation of polarity patterns to the time scale is made more difficult. To confirm correlations to the time scale in those areas, some biostratigraphic control is needed in conjunction with magnetic polarity patterns (Hays and Opdyke 1967; Opdyke et al. 1974; Theyer and Hammond 1974; among others). Biostratigraphic zones are in some cases longer than the paleomagnetic polarity zones for the same time period, but this lack of stratigraphic resolution is usually not critical, because the biostratig-

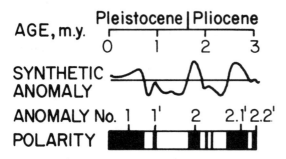

Figure 1.2 *The magnetic anomaly time scale (modified from Ness et al. 1980) for the same time period shown in the magnetostratigraphic time scale in figure 1.1.*

raphy in most cases can determine which portion of the time scale is under study. The polarity patterns may then be correlated to the appropriate reversals in the paleomagnetic or anomaly time scales.

Paleomagnetic Method

The detrital remanent magnetic direction recorded in marine (or terrestrial) sediments is determined by measuring the intensity of magnetization in each of three orthogonal directions in oriented samples of 8 cm^3 which are carefully extracted from the split-face of the core. Many different types of magnetometers are available for paleomagnetic analysis, but each requires very sensitive electronic coils for measuring the weak magnetic intensity of most sediments. After measuring the intensity of magnetization along the three axes, the resultant vector is calculated and expressed as a horizontal component (*declination* with respect to magnetic north) and a vertical component (*inclination* with respect to the horizontal). Declination and inclination are expressed as an angle, in degrees (fig. 1.3).

Standard piston and drill cores of marine sediment are usually unoriented, however, with respect to geographic coordinates. The paleomagnetic orientation of samples from these cores may be determined in a *relative* coordinate system, which is sufficient to determine the paleomagnetic polarity pattern. The standard orientation method for subsamples of a core is to maintain orientation with respect to the split-face of the core and to the vertical. With that technique, fluctuations of the recorded magnetic direction in *relative* horizontal direction or *absolute* vertical direction may be determined. The resulting polarity pattern can then be correlated to the established time scale.

Measurement Problems

The remanent magnetism measured for untreated samples may contain not only the signal imparted at the time of deposition but also a secondary component of magnetization acquired after deposition (Opdyke et al. 1966). This secondary component may

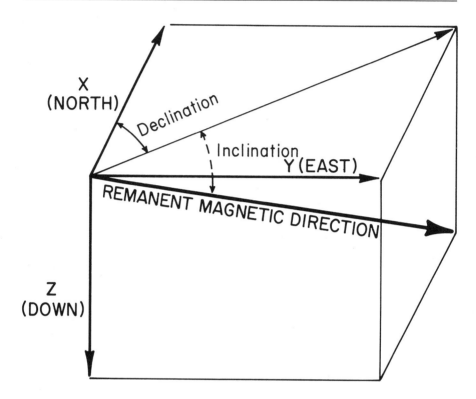

Figure 1.3 *The orientation of the remanent magnetic vector in a sediment sample is expressed as two angles. The declination describes the horizontal and the inclination describes the vertical orientation. The inclination alone may be used to establish the polarity in azimuthally unoriented drill cores.*

be due to postdepositional changes in the sediment (e.g., diagenesis, chemistry, slumping core disturbance, or compaction) or to changes in the magnetic recording that were acquired as the sediment remained in the Earth's magnetic field after deposition. The secondary component of magnetization may obscure the signal from the remanent magnetization acquired at the time of deposition if it is not removed. Fortunately, the secondary component of magnetization in most marine sediments is unstable and may be removed by applying a progressive alternating magnetic field before the remanent magnetization is determined (Opdyke et al. 1966).

The alternating field (AF) demagnetization technique is applied by one of two methods. In both methods, the optimum field for removing the unstable component is chosen from a plot of the remanent magnetization after progressive AF demagnetization. The most common technique, until recently, involved plotting the intensity (J) of magnetization at each demagnetization step as a percentage of the remanent magnetization (J_0) before AF "cleaning" of pilot samples from each lithology in a core. The pattern of J/J_0 decay yields information on the removal of the unstable, secondary component (Opdyke et al. 1966) (fig. 1.4a).

As the unstable component is removed, J/J_0 may (1) decrease quickly to a stable value, indicating a weak secondary component (case 1 in fig. 1.4a); (2) decrease slowly, indicating a stronger secondary component (case 2 in fig. 1.4a), or (3) rise before decreasing owing to removal of a secondary component aligned in a direction opposing the stable component (case 3, fig. 1.4a). By using this method, an arbitrary AF demagnetization level is chosen for removing the secondary component; commonly, this level is $J/J_0 = 0.50$ or the value at which J/J_0 is constant with successive partial demagnetization.

The AF demagnetization technique used most recently involves plotting the demagnetization behavior as vector end-point diagrams (fig. 1.4b) which are projections of the magnetic components on X, Y, and Z axes (Dunlop 1979). The progressive decay of the magnetic components on the vector diagrams (fig. 1.4b) reveals secondary components of magnetization that form a trend that does not approach the origin. After all secondary components are removed with progressive AF "cleaning," the decay trend approaches the origin, and the optimum partial de-

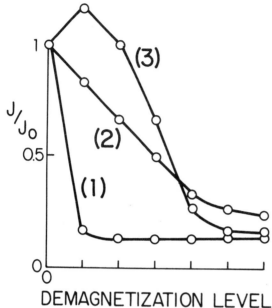

(a)

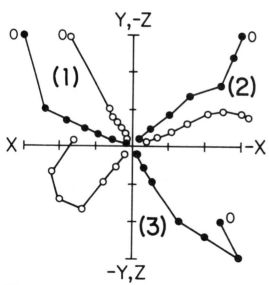

(b)

Figure 1.4 *The secondary, stable direction of magnetization may be removed by partially demagnetizing a sample in progressive steps. Examples for three hypothetical samples are shown for two methods of magnetic "cleaning." (a) The decay of the ratio demagnetized magnetic intensity (J) to original intensity (J_0) may be used to select the appropriate demagnetization level for removing the secondary component. An arbitrary demagnetization level is chosen at $J/J_0 = 0.5$ or at the plateau where J/J_0 does not change with progressive demagnetization. (b) A vector end point diagram of horizontal (closed circles) and vertical (open circles) components of magnetization may be used to detect unstable components (trend away from origin), so that the optimum demagnetization level may be chosen.*

magnetization level is chosen at a value that falls on that trend (fig. 1.4b).

Unfortunately, some marine sediments do not retain a stable component of magnetization, or they possess a component that is too weak to measure with the instruments available. To retain a stable component of magnetization, a sediment sample must contain at least a small amount of magnetite or titanomagnetite with grain sizes less than 10 μm (Haggerty 1970). The amount of magnetic material necessary to carry a stable signal depends on the magnetic susceptibility of the magnetic material, the intensity of the magnetic field, magnetic grain size, instrumentation, and the nature of the dominant sediment type at the site.

In many cases, the right conditions are not met. A common problem involves the secondary coating of grains by chemical precipitation of magnetic minerals presumably carried by pore fluids in the sediment or by oxidation of magnetic minerals after deposition (Kent and Lowrie 1974). The secondary magnetic components may be easily coerced into acquiring a magnetic signal, owing to the small magnetic movement required to re-orient the magnetic domains in the fine-grained precipitates. Therefore, the remanent magnetic direction measured in a sample may change in the space of seconds while the sample sits in the Earth's ambient field in the laboratory or as the sample is extracted from or inserted into the measuring instrument. The stability of the magnetic recording may be tested by remeasuring the remanent magnetization of the same sample after placing the sample in different orientations while storing it in the laboratory. If stability tests reveal fluctuations in remanent direction with reorientation, those samples with an unstable magnetic recording must be excluded from consideration.

Polarity Determination

Either the declination or inclination of the remanent magnetization (fig. 1.3) may be used to determine the magnetic polarity of sediment samples from a core or section in which the samples are oriented with respect to geographic coordinates. The technique is demonstrated in the hypothetical section shown in figure 1.5. The declination points north, and the inclination is positive

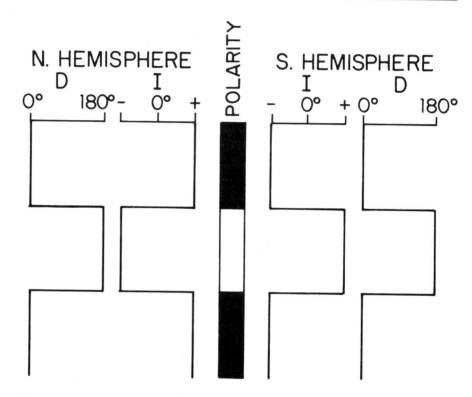

Figure 1.5 *The declination (D) and inclination (I) (see fig. 1.3) of a marine sedimentary section may be used to determine the magnetic polarity. The pattern of shifts in declination and inclination are different in the Northern and Southern Hemispheres.*

(downward) for sediment deposited in the Northern Hemisphere during a period of normal (like today) polarity. An opposite orientation for the inclination (i.e., upward) is observed during normal polarity in the Southern Hemisphere (fig. 1.5). Sediment deposited during a period of reversed polarity will have a southern declination and negative (upward) inclination in the Northern Hemisphere, and a positive (downward) inclination in the Southern Hemisphere.

Because inclination is measured with respect to the vertical direction only (fig. 1.3), it may be determined in cores with only bottom-top integrity. As a result, a positive (downward) inclination in cores from the Northern Hemisphere is interpreted as *normal* polarity, and a negative inclination is considered *reversed* polarity. Therefore, a polarity log of any sedimentary core may be determined in spite of the lack of orientation in geographic coordinates, and the resulting polarity pattern may be correlated easily to the magnetic time scale (fig. 1.1) unless unconformities are present in the section. An example of the correlation of sedimentary sections recovered in unoriented piston cores from the South Atlantic to the paleomagnetic time scale is shown in figure 1.6; a diatom biostratigraphic zonation was used to assign the correlation of magnetic reversals (Ledbetter and Ciesielski 1982). In this example, the combination of biostratigraphy and magnetostratigraphy revealed hiatuses in the sedimentary section that would have been difficult to define accurately with a single stratigraphic method.

Applications of Pleistocene Magnetostratigraphy

Magnetostratigraphy may be used directly or indirectly to assign ages to a sedimentary section. With the direct method, the magnetostratigraphy is determined for the section under study, whereas with the indirect method, the magnetostratigraphy is used to develop an age model for other stratigraphic horizons that are found in the studied section. In the latter case, the age of the dated horizons is assigned by interpolating between the ages of the reversal boundaries (fig. 1.1). This indirect method of assigning ages has proved to be of major significance in marine stratigraphy, and examples are discussed below.

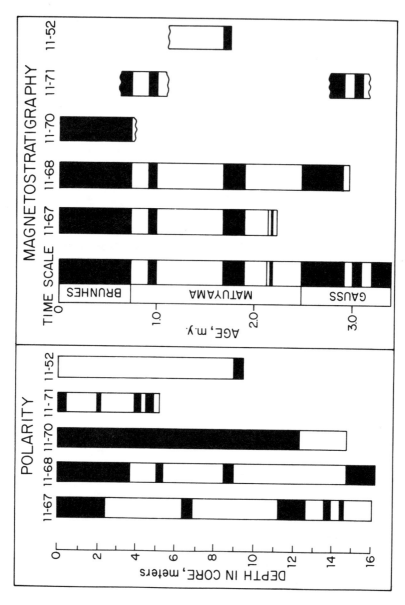

Figure 1.6 *The paleomagnetic polarity of piston cores (left) may be correlated to the magnetostratigraphic time scale (right) with the use of biostratigraphy. In this example, a diatom biostratigraphy was used for the correlation of sedimentary sections recovered by piston cores that contained hiatuses and large differences in sedimentation rates (from Ledbetter and Ciesielski 1982).*

Age Assignment to Stratigraphic Horizons

Because magnetostratigraphy provides globally synchronous ages of magnetic reversal boundaries, the method may be used to date stratigraphic horizons traced into sections with a reliable magnetostratigraphy. The synchroneity of those horizons may be verified in other widely separated sections by tracing the horizons into another site where a magnetostratigraphic record is available. Once verified, the age of the stratigraphic horizon may be extrapolated to other sections without determining the magnetostratigraphy again. This method is particularly useful in establishing the age of biostratigraphic horizons for all fossil groups, regardless of regional or interocean differences in paleoenvironment (for example, Berggren and van Couvering 1974).

The absolute age of first and last appearance datums of the major fossil groups have been established in all ocean basins by correlation to deep-sea cores with a magnetostratigraphy, and correlated to boundary-stratotype and type sections by using biostratigraphy. Unfortunately, many of the boundary-stratotype and type sections cannot be dated directly with magnetostratigraphy, owing to lack of a stable magnetic recording (for example, Watkins et al. 1974). The age of Pleistocene microfossil datums for the major fossil groups has been summarized by Berggren et al. (1980) and is presented in figure 2.1. The age of the Pliocene/Pleistocene boundary, as defined in the Gulf of Mexico, is also based on the correlation of biostratigraphic horizons to magnetostratigraphy (fig. 2.2), as is the boundary in the open ocean (fig. 2.3). Naturally, once the age of a biostratigraphic datum is established for a basin, it may be extrapolated to other sections without going through the task of determining the magnetostratigraphy in those sections.

In addition to biostratigraphic datums, other stratigraphic horizons may be dated with magnetostratigraphy. The most notable of these is the Pleistocene oxygen isotope stratigraphy. The ages of oxygen isotope horizons older than a radiometrically determined age of 127 thousand years before present (KYBP) (Broecker and Van Donk 1970) were assigned correctly only after the oxygen isotope stratigraphy was determined in piston core V28–238 (fig. 3.6) in which a magnetostratigraphy was available (Shackleton and Opdyke 1973). Until the oxygen isotope stratigraphy of that core was completed, a heated controversy existed on the timing of isotope fluctuations (Emiliani and Shackleton 1974). In core V28–238 (and other cores), oxygen isotope stage 19 falls near the Brunhes/Matuyama boundary at 720 KYBP (fig. 1.1). The age of fluctuations in the oxygen isotope stratigraphy between 127 and 720 KYBP was originally established by interpolation from the magnetostratigraphy (Shackleton and Opdyke 1973). Oxygen isotope records older than the Brunhes record of V28–238 are dated solely on the basis of magnetostratigraphy (fig. 3.11).

Advent of the Hydraulic Piston Corer

Magnetostratigraphy of early Pleistocene sediments with high sedimentation rates that are beyond the recovery limit of standard piston cores has been difficult, owing to extreme disturbance caused by rotary drilling on the *Glomar Challenger,* which permitted recovery of a longer stratigraphic record. The paleomagnetic polarity of Pleistocene sediments recovered by the *Glomar Challenger* may now be determined, however, in very thick marine sections comparable to the sedimentary sections in the Gulf of Mexico. This new advance in magnetostratigraphy was made possible by the invention of the hydraulic piston corer (HPC), a tool that recovers nearly complete, undisturbed sediment throughout a thick sedimentary sequence. The HPC consists of a 9.5m core barrel, which is extruded in front of the drill bit (fig. 1.7) by back-pressuring the barrel with seawater until shear pins fail and release the core barrel, which slides past a piston and into the sediment (Prell, Gardner et al. 1980). After the HPC has fully extended, the tool is retrieved and a new HPC is sent down the hole after the drill bit is lowered through sediment already cored.

The HPC fully "strokes out" and recovers a 9.5m, undisturbed sediment core, until encountering sediment with a shear strength of 1200 g/cm^2. Full penetration is not possible in sediment with a shear strength of 1200 g/cm^2, but undisturbed sediment is recovered in shorter cores.

The first magnetostratigraphy in HPC cores was obtained for DSDP Leg 68 (Kent and Spariosu 1982). To obtain a complete section for magnetostratigraphy, the HPC must core the entire sedimentary section present at the site. Although uncomformities may be present at the site, the polarity pattern and biostratigraphy may be used to detect a hiatus (fig. 1.6).

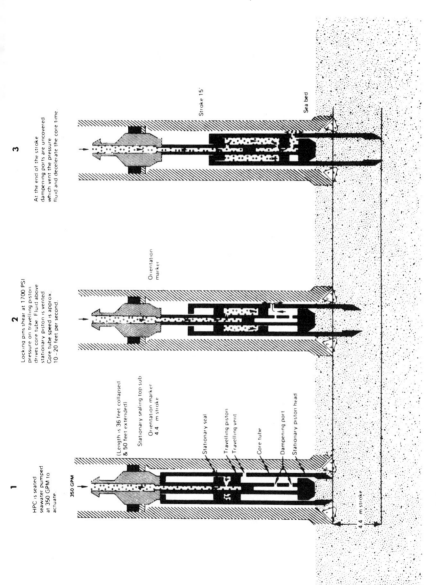

Figure 1.7 *The hydraulic piston core is a downhole tool developed by DSDP which may be used to recover undisturbed sediment which is suitable for magnetostratigraphy in a rotary-drilled section. A 9.5m core is extruded past a piston in front of the drill bit in order to recover a core.*

Four holes were cored with the HPC at one location in the Columbian Basin (western Caribbean), and the magnetostratigraphy was determined with the aid of biostratigraphic zonations (fig. 1.8). The relatively high accumulation rates (3 cm/10^3 yr.) allowed excellent resolution of the paleomagnetic polarity pattern, so ages could be placed at specific horizons in the last 5.0 MY of sediment (145m subbottom) (fig. 1.8). The magnetostratigraphy in HPC cores provides a first-order stratigraphy for soft (high–water content) sediments within the Pleistocene that could be dated previously in rotary-drilled cores only with biostratigraphy.

Pleistocene Sections in HPC Cores

The HPC is a relatively new tool on the *Glomar Challenger,* and only a few magnetostratigraphic records of Pleistocene sections are published at this time. The high-resolution records of the Pleistocene obtained on two of the first DSDP Legs (68 and 71) to use the HPC allowed further resolution of the age of biostratigraphic zonal boundaries. The magnetostratigraphy of the Pleistocene section at Site 502 in the western Caribbean is shown in figure 1.8. The good preservation of calcareous microfossils at this site allowed placement of biostratigraphic zonal boundaries for early Pliocene to Holocene, for both foraminifers and calcareous nannofossils within the magnetostratigraphy (Prell and Gardner 1982) (fig. 1.9). The stratigraphy reveals that a complete section was cored at Site 502, and the age of faunal and oxygen isotope (fig. 3.11) boundaries was determined with greater resolution than was previously possible (Keigwin 1982).

Another high-resolution record of magnetostratigraphy of the Pleistocene was obtained at Site 503 in the eastern equatorial Pacific Ocean (Prell and Gardner 1982). The paleomagnetic polarity could be identified from early Pliocene to Holocene, and the biostratigraphic zones for foraminifers, calcareous nannofossils, radiolaria, and diatoms were placed within a magnetostratigraphic framework (fig. 1.10). The greater resolution in a continuous section provided by the HPC allows a concomitantly greater resolution of biostratigraphic zonal boundaries at Site 503 (Keigwin 1982).

A third high-resolution record of a Pleistocene/Pliocene section was obtained with the HPC at Site 514 in the southwest Atlantic Ocean. The paleomagnetic polarity at Site 514 was

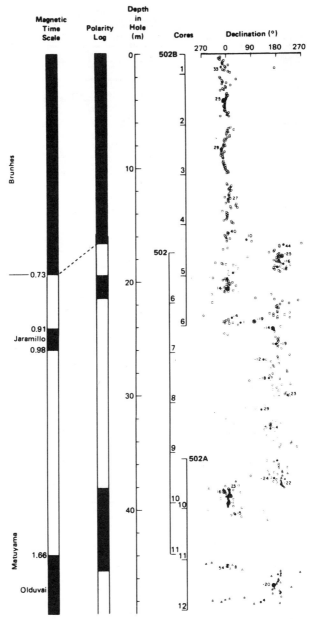

Figure 1.8 *The magnetostratigraphy (left) of hydraulically piston-cored DSDP Site 502 (Holes 501, 502, 502B) is interpreted from the declination (see fig. 1.3) direction. A declination of 0° corresponds to normal-polarity and 180° to reversed-polarity sediments (from Kent and Spariosu 1982).*

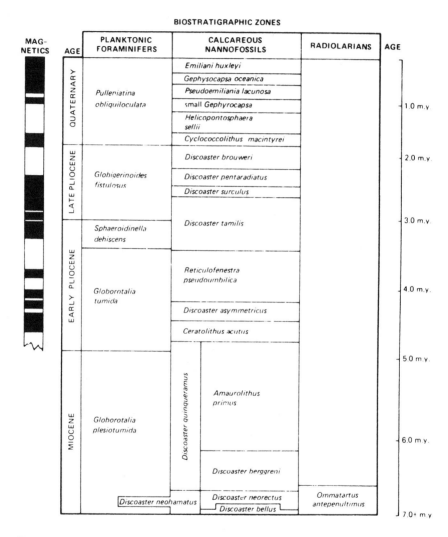

Figure 1.9 *The magnetostratigraphy at HPC Site 502 (see fig. 1.8) may be used to assign ages to biostratigraphic zonal boundaries. (Modified from Prell and Gardner 1982.)*

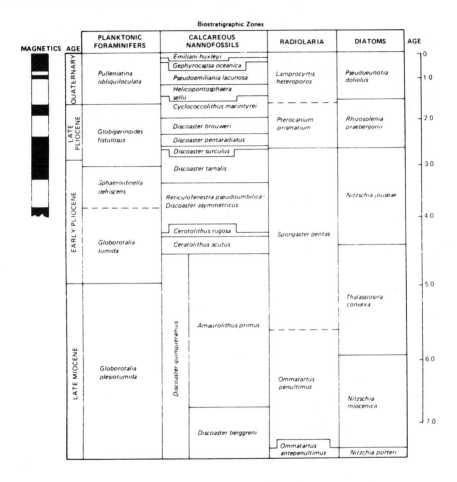

Figure 1.10 *The magnetostratigraphy at HPC Site 503 may be used to assign ages to biostratigraphic zonal boundaries. (Modified from Prell and Gardner 1982.)*

correlated to the magnetostratigraphy by using a diatom and silicoflagellate zonation (fig. 1.11) (Salloway 1983). The high-resolution record at this site, particularly in the late Pliocene, allows a high-resolution record of biostratigraphic zonal boundaries and lithofacies fluctuations.

The acquisition of these continuous HPC sections of undisturbed core is an exciting development in Pleistocene stratigraphy. Instead of piecing together parts of the record from widely separated, low-resolution records, it is now possible to retrieve a magnetostratigraphy from one site in an area with high sedimentation rates. Therefore, the greater age resolution of faunal, floral, isotope, and lithologic horizons will add a new dimension to the study of Pleistocene marine sections.

Possible Uses in Petroleum Exploration

For many years, magnetostratigraphy has been used in the petroleum industry to resolve stratigraphic and structural problems, but the technique is usually applied to problems involving terrestrial exposures of marine sections. The lack of use of magnetostratigraphy in offshore exploration is principally due to the previous inability of geologists to retrieve undisturbed sedimentary sections when rotary-drilling. With the advent of the HPC, however, it is now possible to overcome the problem of sediment disturbance, possibly by developing a sidewall corer that could retrieve samples with vertical orientation. The major problem remaining at this time is to convince industry of the benefits of adding another stratigraphic tool in their exploration techniques.

The HPC technology was a routine part of the scientific program on *Glomar Challenger* for Legs 68–96. The HPC was used principally to recover sediment in areas with low sedimentation rates (1–10 cm/1000 yr.). In those areas, the HPC must be used to recover continuous sections in order to acquire the needed stratigraphic resolution. In the northern Gulf of Mexico, however, the high accumulation rate makes it unnecessary to retrieve a continuous core. A 9.5m HPC core every 50–100m of drilled section would be sufficient to define major polarity patterns (fig. 1.1). The magnetostratigraphy may be determined within less than an hour after recovery of sediment, by developing highly portable paleomagnetic equipment on the drilling rig or ship. The minimal cost in time for taking a few cores to obtain the magnetostratigraphy in a stratigraphic well may be very cost ef-

Figure 1.11 *The magnetostratigraphy at HPC Site 514. (Modified from Salloway 1983.)*

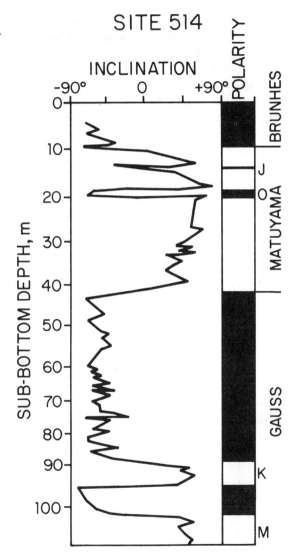

fective. The major problem to be encountered would be the shear strength of deep sediments, which would prevent penetration of the corer. However, spot-coring by a very short version of the HPC or oriented sidewall cores will yield satisfactory resolution of the magnetostratigraphy.

Magnetostratigraphy also may be used for other purposes in stratigraphic studies or basin analysis. Because the polarity reversal pattern (fig. 1.1) is so distinctive, it may be possible to identify repeated sections in tectonically active regions. With a minimum of biostratigraphic control, the magnetostratigrapher may be able to delineate sections where the polarity pattern indicates fault-repeated sections. Additionally, the magnetostratigraphy may be used to date seismic horizons and/or unconformities in a sedimentary section. The procedure is used routinely in deep-sea cores to obtain an absolute age on widespread unconformities and regional seismic horizons in order to compile a regional synthesis of a depositional basin (fig. 5.6).

Summary

The magnetostratigraphy of marine sediments may be used to provide absolute age determinations at four time-stratigraphic horizons (0.72, 0.91, 0.98, and 1.66 MYBP) within the Pleistocene. These horizons may be used as direct stratigraphic markers or as means of indirectly assigning ages to micropaleontologic, lithologic, or oxygen isotopic fluctuations that occur between paleomagnetic reversal boundaries. No other stratigraphic method exists that can provide a globally synchronous time stratigraphy in marine sediments regardless of depositional environment.

The hydraulic piston corer developed by DSDP and fully tested for the first time on Leg 68 (1979) may be used to obtain a long, undisturbed core suitable for magnetostratigraphy in a shallow rotary-drilled section where compaction does not prevent penetration. Magnetostratigraphy in thick Pleistocene sedimentary sequences, as in the Gulf of Mexico may be determined by only spot-coring the sections to identify the major polarity boundaries in Pleistocene or older sediment. Magnetostratigraphy also may be used to identify repeated sections in tectonically active areas and to date seismic horizons and/or unconformities in a regional analysis.

2

Pleistocene Planktonic Foraminiferal Biostratigraphy and Paleoclimatology of the Gulf of Mexico

Robert Thunell
Department of Geology
University of South Carolina
Columbia, South Carolina   29208

2

Introduction

The large number of relatively continuous sedimentary sections recovered through deep-sea drilling during the last decade has resulted in a greatly improved understanding of the stratigraphic ranges of members of the various microfossil groups. This enhanced stratigraphic resolution, combined with a more precise absolute time scale, has led to the establishment of highly refined biochronologic frameworks for the Jurassic (Van Hinte 1976a), Cretaceous (Van Hinte 1976b), and Cenozoic (Hardenbol and Berggren 1978; Berggren and Van Couvering 1974; Berggren et al. 1983). In principle, biochronology relies on irreversible evolutionary events or datums, such as first and last appearances, to subdivide geologic time. Planktonic foraminiferal evolution during the Cenozoic has proceeded at an average rate of about 1.5 new appearances per million years and 1.3 extinctions per million years (Thunell 1981a). Utilization of these various datums provides the basis for different biochronologic schemes in which the last 65 million years have been organized into a series of units, each of which is on the order of 1 to 2 million years in duration.

Although the duration of these zones provides adequate biostratigraphic resolution for the Cenozoic, it leads to a problem when one considers only the most recent geologic epoch, the Pleistocene. If the Pleistocene is only 1.6 to 1.8 million years in duration, one would not expect many, if any, of these evolutionary planktonic foraminiferal datums. Indeed, the first appearance of *Globorotalia truncatulinoides* and the last appearance of *Globigerinoides obliquus*, which are often used to identify the Pliocene/Pleistocene boundary, are probably the two most recent evolutionary events that have occurred in planktonic foraminifera that can be used on a global basis. As a result, further subdivision of the Pleistocene must rely on additional findings of microfossil groups and "nonevolutionary" biostratigraphic events that may only be of local or regional significance. Examples of "nonevolutionary" biostratigraphic events and the geographic regions in which they have been utilized are given in table 2.1.

In most instances, these represent migrational events or a response to changing oceanographic/climatic conditions. The more important evolutionary datums for the different microfossil groups that have been used to subdivide the Pleistocene have

Event	Age (YBP)	Region
LAD G. *menardii flexuosa*	~90,000	Gulf of Mexico
LAD G. *hexagona*	~90,000	Gulf of Mexico
LAD G. *ruber* (pink)	~120,000	Indo-Pacific
LAD G. *pseudofoliata*	~220,000	Indo-Pacific
LAD G. *tosaensis*	~590,000	Indo-Pacific
LAD G. *tosaensis*	?	Gulf of Mexico
FAD G. *conglomerata*	~610,000	Indo-Pacific
FAD G. *ruber* (pink)	?	Gulf of Mexico
FAD P. *finalis*	?	Gulf of Mexico
FAD P. *finalis*	~1,300,000	Indo-Pacific

Table 2.1 *Examples of Pleistocene "Nonevolutionary" or Regional Planktonic Foraminiferal Events*

been synthesized recently by Berggren and others (1980) and are summarized in figure 2.1.

In this review, different qualitative and quantitative approaches that commonly have been used to zone biostratigraphically the Pleistocene of the Gulf of Mexico will be evaluated. In addition, the way in which these planktonic foraminiferal studies have been used to reconstruct the Pleistocene paleoclimatic history of the Gulf of Mexico will be discussed.

Pliocene/Pleistocene Boundary: Age, Definition, and Recognition

A great deal of confusion and controversy still surrounds the identification of the Pliocene/Pleistocene boundary, owing in large part to fundamental differences in stratigraphic philosophy (see Berggren and Van Couvering 1974; Pelosio et al. 1980; and Colalongo et al. 1982 for a review). Historically, many workers have recognized, and continue to recognize the Pliocene/ Pleistocene boundary on the basis of climatic criteria. This approach evolved out of Forbes's (1846) association of the onset of climatic deterioration with the beginning of the Pleistocene.

In principle, a time-stratigraphic boundary should be defined by a unique physical reference point—the "boundary stratotype," or the "golden spike." All time-stratigraphic boundaries should be characterized by a precise definition that means the same thing to all workers everywhere (Hedberg 1976). Because the Pliocene/ Pleistocene boundary is a time horizon or chronohorizon, it should be considered a universally standard reference point. As proposed at the Seventh INQUA Congress in Denver (1965), the stratotype for the Pliocene/Pleistocene boundary is the marker bed (Bed G–G′) of Gignoux (1910 and 1913) at Le Castella, Italy. This boundary has no intrinsic climatic or paleontologic significance, although such criteria can be employed in its global recognition and correlation. As was succinctly stated by Berggren and Van Couvering (1974, p. 86), "chronostratigraphic boundaries are not determined by convenience."

The fossil record provides the best means for the recognition and correlation of chronostratigraphic boundaries (Berggren and Van Couvering 1974). Biostratigraphic correlation of the bound-

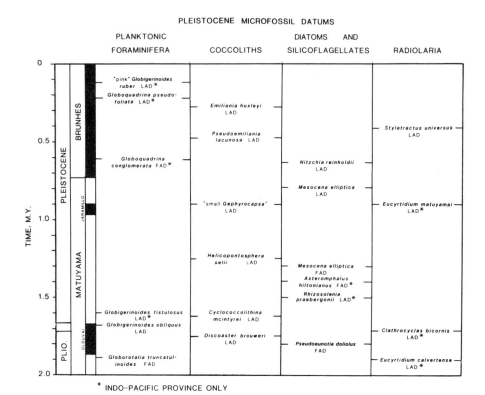

Figure 2.1 *Correlation of Pleistocene microfossil datums with the paleomagnetic time scale (based on-Berggren and others, 1980). The time scale is from Mankinen and Dalrymple (1979). LAD stands for last appearance datum. FAD stands for first appearance datum. Biostratigraphic datums marked with an asterisk are restricted to the Indo-Pacific province.*

ary-stratotype section at Le Castella with paleomagnetically dated deep-sea cores indicates that the Pliocene/Pleistocene boundary is equivalent to the top of the Olduvai Event (Haq et al. 1977). This results in an age of about 1.6 to 1.8 million years for the Pliocene/Pleistocene boundary, depending on which polarity time scale is used (Cox 1969; McDougall 1977; Mankinen and Dalrymple 1979) (see chap. 1 for a review of the development of the polarity time scale). Biostratigraphically, the first appearance of *Globorotalia truncatulinoides* occurs in the latest Pliocene in association with the base of the Olduvai Event, whereas the last appearance of *Globigerinoides obliquus* occurs in the upper part of the Olduvai Event near the Pliocene/ Pleistocene boundary (Haq et al. 1977; Rio et al., in press).

More recently, Selli and others (1977) proposed that a new boundary stratotype at the Vrica section in Calabria, Italy be adopted, because of better stratigraphic continuity than that at the Le Castella section. Radiometric dating of ash layers from within the Vrica section yield an age of less than 2 million years for the Pliocene/Pleistocene boundary at this locality (Obradovich et al. 1982). In addition, recent paleomagnetic (Tauxe et al. 1983) and biostratigraphic (Backman et al. 1983) studies indicate that the boundary level at Vrica reveals an age of approximately 1.6 million years. As a result, the boundaries as recognized at both Vrica and Le Castella are time equivalent.

An age of less than 2.0 million years for the Pliocene/ Pleistocene boundary disassociates it clearly with the initiation of major Northern Hemisphere glaciation. A substantial amount of evidence now indicates that major ice-sheets developed in North America and Europe from 2.5 to 3.0 million years ago (Curry 1966; Berggren 1972; McDougall and Wensink 1966; Shackleton and Opdyke 1977; Backman 1979; Prell 1983; Shackleton and Cita 1979; Shackleton et al. 1982; Thunell and Williams 1983). In addition, investigations of the boundary stratotype section at Le Castella show no evidence of a major climatic change across the Pliocene/Pleistocene boundary (Emiliani et al. 1961; Selli 1967). Unfortunately, the concept that the Pleistocene is synonymous with glaciation is still used and has led to misidentification and miscorrelation of sediments presumed to be of Pleistocene age. In this sense, Beard and Lamb (1968), Beard (1969) and Beard and others (1982) have conceptually correlated the base of the Pleistocene in the Gulf of Mexico with the first indication of severe climatic deterioration or the beginning of the Nebraskan glacial stage, and have associated it with the top of the Kaena paleomagnetic event, which has been dated at approximately 2.9 million years (McDougall 1977; Mankinen and Dalrymple 1979). This results in the unacceptable situation of using a Pliocene/Pleistocene boundary in the Gulf of Mexico that is more than 1 million years older than that determined for either of the two proposed stratotype sections.

A number of different planktonic foraminiferal biostratigraphic criteria have been used to identify the Pliocene/Pleistocene boundary in both the Gulf of Mexico and the open ocean. In the Gulf of Mexico, Lamb and Beard (1972) have used the last ap-pearances of *Globoquadrina altispira* and *Globoquadrina venezuelana* to define the base of the *Globorotalia tosaensis* zone, the lower limit of which they correlate with the Pliocene/Pleistocene boundary (fig. 2.2). If these biostratigraphic events are truly evolutionary datums, they should serve as isochronous horizons for recognition of the Pliocene/Pleistocene boundary on a global basis. Correlation of these biostratigraphic events from one ocean basin to another, coupled with their calibration to the paleomagnetic time scale indeed demonstrates that most of these events can be considered evolutionary datums (Ryan et al. 1974; Saito et al. 1975; Berggren 1977; Thunell 1981b). However, these studies also indicate that many of the datums that have been used by Lamb and Beard (1972) and Beard and others (1982) to identify the Pliocene/Pleistocene boundary in the Gulf of Mexico actually occur in the late Pliocene (fig. 2.3). As a result, it is not surprising that the planktonic foraminiferal datums used by Beard and Lamb (1968), Beard (1969), and Lamb (1969) to identify the Pliocene/Pleistocene boundary in the Gulf of Mexico do not mark the boundary at either the Le Castella or Vrica sections.

The apparent discrepancy between the Gulf of Mexico and the open-ocean record appears to have been resolved in the recent work of Brunner and Keigwin (1981). In their study of a core drilled from the DeSoto Canyon it was demonstrated that the ranges of *Globorotalia miocenica* and *Globorotalia pertenuis* do not overlap with those of *G. truncatulinoides* and *Pulleniatina obli-quiloculata* as previously reported by Lamb and Beard (1972).

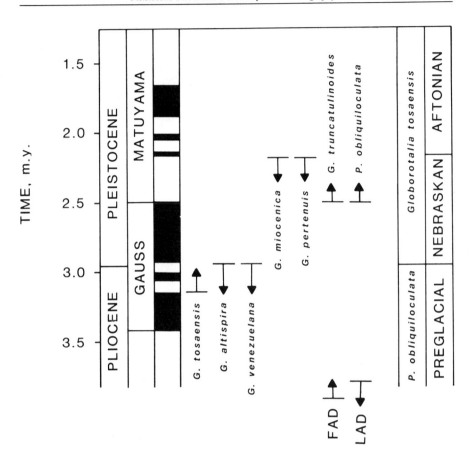

Figure 2.2 *Planktonic foraminiferal biostratigraphic recognition of the Pliocene/Pleistocene boundary in the Gulf of Mexico as proposed by Lamb and Beard (1972), and its relationship to the paleomagnetic time scale and North American glacial stages.*

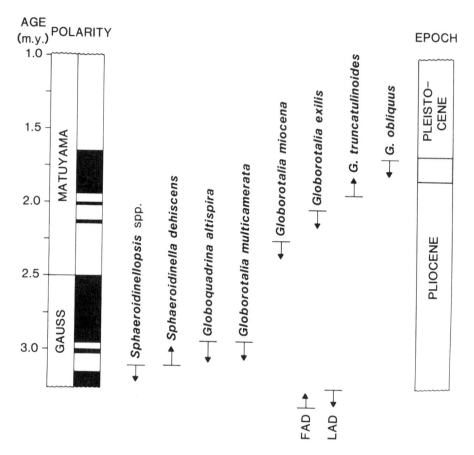

Figure 2.3 *Planktonic foraminiferal biostratigraphic criteria used to identify the Pliocene/Pleistocene boundary in the open ocean and its relationship to the paleomagnetic time scale (based on Haq et al. 1977).*

Brunner and Keigwin (1981) used the first appearance of *G. truncatulinoides* to approximate the Pliocene/Pleistocene boundary and placed the last appearances of *G. altispera, G. venezuelana, G. miocenica* and *G. pertenuis* in the Pliocene, in agreement with the open-ocean record. In this manner, the boundary is identified purely in terms of stratigraphic criteria, and the concept of climatic change plays no part in its recognition.

Qualitative Biostratigraphy

The Presence-Absence Approach

The lack of a sufficient number of evolutionary datums in the Pleistocene has led micropaleontologists to look for different criteria to subdivide the Pleistocene stratigraphically. One commonly used approach is simply based on the presence or absence of key planktonic foraminiferal species. In most instances, the presence-absence pattern of a species is due to climatically related migrations, and this approach therefore provides some basic paleoenvironmental information, together with the desired biostratigraphic results.

The most commonly used example of this presence-absence approach is the so-called *Globorotalia menardii* zonation established by Ericson and Wollin (1968). In its simplest form, this zonation allows for a tenfold subdivision (zones Q–Z) of the Pleistocene (fig. 2.4). Ericson and Wollin (1968) considered intervals in which *G. menardii* is present to be interglacials, whereas its absence denotes glacial periods. This zonation is only applicable within the Atlantic, since *G. menardii* is present during both interglacials and glacials in the Indo-Pacific region. These long-term glacial-interglacial cycles are the result of global climatic changes, and the timing of these faunal zones is generally considered to be synchronous from one region of the Atlantic to another. However, Berger (1982) has suggested that this may not always be the case. For example, the reappearance of *G. menardii* at the Y/Z boundary occurs slightly earlier in the equatorial Atlantic than in the Gulf of Mexico or off northwestern Africa. Similarly, Kennett and Huddleston (1972a) reported that the X/Y boundary is diachronous between the Gulf and the equatorial Atlantic, with this boundary being older in the Gulf of Mexico.

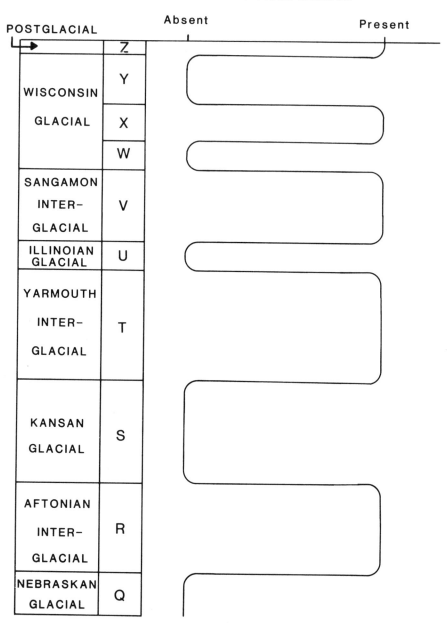

Figure 2.4 *Biostratigraphic zonation of the Pleistocene based on the presence/absence of the planktonic foraminifera* Globorotalia menardii, *and its proposed relationship to the North American glacial stages (based on Ericson and Wollin 1968).*

On the basis of inferred climatic significance of the presence or absence of G. *menardii,* the different zones have been correlated in a general way with the classical North American Pleistocene stages (fig. 2.4). An in-depth evaluation of the problems involved in correlating Pleistocene marine and terrestrial climatic records is presented in Fillon (this volume, chap. 5) and Kukla (1977).

The "G. *menardii* zonation" thus allows for a rapid and relatively reliable subdivision of the Pleistocene. However, the stratigraphic resolution provided by this zonation decreases with increasing age, owing to the fact that the older zones incorporate longer intervals of time (table 2.2). The four most recent zones (W–Z) have an average duration of about 50,000 years, whereas zones R through V each span an average of almost 300,000 years. When used in the strictest sense, this zonation is inadequate for high-resolution studies of pre–late Pleistocene sequences. This biostratigraphic scheme allows for the identification of five broad "glacial" periods (zones Q, S, U, W, and Y). However, on the basis of faunal and isotope studies of deep-sea sequences, it is known that many more glacial episodes occurred during the Pleistocene (Van Donk 1976; Shackleton and Opdyke 1973 and 1976), with glacial-interglacial cycles typically showing a periodicity of about 100,000 years (Hays et al. 1976). As a result, these faunal zones should not be interpreted as being simply glacial or interglacial. More accurately, most of these zones represent both glacial and interglacial conditions. For example, "glacial" zone Y actually incorporates isotopically defined interglacial stage 3 along with glacial stages 2 and 4 (see chap. 3 for a discussion of late Pleistocene glacial-interglacial isotope stages).

When originally defined (Ericson and Wollin 1968), the zones in this biostratigraphic scheme were based purely on the presence or absence of G. *menardii.* Since then, this zonation has been utilized by many investigators, and, often, the criteria for recognizing the various zones has been modified. Specifically, the simple presence-absence definition often has been dropped in favor of something more quantitative (i.e., differentiating between its presence in high or low abundances). This obviously can lead to some confusion, because a particular zonal boundary could be placed at varying levels within a sequence, depending on which set of criteria are used. Within the Gulf of Mexico, Emiliani et al.

A.   Ages for Zonal Boundaries

Zonal-Boundary	Isotope Stage	Age (YBP)
Z/Y	1/2 Boundary	11,000
X/Y	late Stage 5	90,000
W/X	5/6 Boundary	128,000
V/W	6/7 Boundary	195,000
U/V	mid Stage 13	485,000
T/U	mid Stage 16	610,000
S/T		990,000
R/S		1,300,000
Q/R		1,650,000

D.   Duration of zones

Zone	Duration (years)
Z	11,000
Y	79,000
X	38,000
W	67,000
V	290,000
U	125,000
T	380,000
S	310,000
R	350,000

Table 2.2 *Pleistocene Chronology*

(1975) and Brunner and Keigwin (1981) adhered to the strict presence-absence format in identifying these zones. Alternatively, Kennett and Huddleston (1972a) used G. *menardii* abundance criteria as well as several other faunal indices to define these zones in their study of piston cores from the Gulf of Mexico. This latter approach will not lead necessarily to the placement of zonal boundaries at levels equivalent to those selected by using the pure Ericson and Wollin definition. For example, using a strict presence-absence approach will result in the base of an "interglacial" zone being placed lower in the section than would be true if a specific minimum abundance level of G. *menardii* were used. In deeper parts of the Gulf, where sedimentation rates are low, the different approaches will result in relatively small discrepancies. However, the problem is more acute in marginal regions where sedimentation rates are very high. In this situation, there could be a significant difference in the stratigraphic placement of a boundary, depending on which approach is used.

For most stratigraphic studies, it is desirable to be able to place the identified stratigraphic units within some type of time framework. As has already been discussed, the calibration of several biostratigraphic events to the paleomagnetic time scale has produced absolute-age estimates for the Pliocene/Pleistocene boundary (i.e., Haq et al. 1977; Backman et al. 1983; Tauxe et al. 1983; see also chap. 1). Several attempts also have been made to impart a better time perspective to the G. *menardii* Zones. Perhaps the best example of this is the faunal, isotopic, and paleomagnetic study of Van Donk (1976) on an Atlantic piston core that contains a relatively complete Pleistocene record. The paleomagnetic stratigraphy provides the basic time control for dating the zonal boundaries. In conjunction with this, well-established oxygen isotope chronology (Shackleton and Opdyke 1976; see also chap. 3) can be used to refine these age estimates. For example, it has been well established that the Y/Z boundary is associated with Termination I (isotope stage 1/2 boundary) with an age of about 11,000 years (Broecker 1966; Broecker and Ku 1969; Broecker and Van Donk 1970; Shackleton and Opdyke 1973 and 1976). Age estimates for each of the zonal boundaries, and their relationship with respect to the oxygen isotope stages as reported in Van Donk (1976) are given in table 2.2.

Nonevolutionary or Regional Biostratigraphic Events

In addition to the "G. *menardii* Zonation," there are a number of biostratigraphic events that either are nonevolutionary in nature or have restricted geographic usefulness that provide additional qualitative means for subdividing the Pleistocene (table 2.1). Three of these datums—the first appearance of *Globoquadrina conglomerata,* the last occurrence of *Globoquadrina pseudofoliata,* and the last occurrence of pink-pigmented *Globigerinoides ruber* have been dated at approximateley 610 KYBP, 220 KYBP and 120 KYBP, respectively (Thompson and Sciarillo 1978; Thompson et al. 1979). However, these "datums" are useful only in the Indo-Pacific region.

A number of similar planktonic foraminiferal events have been used by numerous workers in the Gulf of Mexico. Two of the most commonly used of these events are the last appearances of G. *menardii flexuosa* and *Globoquadrina hexagona.* These are not true evolutionary last appearances, because G. *menardii flexuosa* is still extant in the Indian Ocean (Bé 1970), and G. *hexagona* is present in the Pacific (Parker 1967). According to Kennett and Huddleston (1972b), these last appearances occur in close proximity to each other in the vicinity of the X/Y boundary, and therefore fall within the upper part of isotope stage 5 (see chap. 3).

Several other less well-defined Pleistocene "datums" that have been used in the Gulf include the first appearance of pink-pigmented G. *ruber,* the first appearance of *Pulleniatina finalis,* and the last appearance of G. *tosaensis.* According to Lamb and Beard (1972) both the "P. *finalis* datum" and the "G. *ruber* datum" occur during the Illinoian glacial, which, in turn, has been correlated to zone U (fig. 2.4) (Ericson and Wollin 1968). The reliability of these correlations and the usefulness of the datums is somewhat uncertain. For example, Brunner and Keigwin (1981) identified the first appearance of P. *finalis* within a Gulf unit they assigned to zone V of Ericson and Wollin (1968). In the Pacific, Hays and others (1969) placed the first appearance of P. *finalis* midway between the Jaramillo and Olduvai Events, thus giving it an age from about 1.0 to 1.7 million years. Since zone V occurs during the Brunhes Epoch (van Donk 1976), it is apparent that the "P. *finalis* datum" is substantially older in the Pacific than it is in the Gulf of Mexico.

The last appearance of G. *tosaensis*, as it is used in the Gulf of Mexico, defines the top of the early Pleistocene G. *tosaensis* zone (Lamb and Beard 1972). In addition, Beard et al. (1982) correlated this event with the early Pleistocene Gilsa normal polarity event (see chap. 1). However, in the Pacific, the last occurrence of G. *tosaensis* occurs just above the base of the Brunhes Epoch, at approximately 590 KYBP (Thompson and Sciarillo 1978). This again illustrates the diachroneity of these biostratigraphic events from one region to another. What is definitely needed within the Gulf of Mexico is a detailed paleomagnetic and oxygen isotope study of Pleistocene sections, so that first-order correlation can be made with all of these "nonevolutionary" or regional biostratigraphic events. This would greatly enhance the time significance and usefulness of these "datums."

Quantitative Planktonic Foraminiferal Biostratigraphy

A number of quantitative approaches can be used to improve Pleistocene biostratigraphic resolution. This section will discuss three of these approaches and how each has been used in the Gulf of Mexico. These include (1) single-species abundances, (2) coiling ratios, and (3) total faunal abundances.

Single-Species Abundances

Variation in the abundance of individual species of planktonic foraminifera during the Pleistocene is primarily a response to climatically controlled changes in surface-water conditions (temperature and salinity). In addition to the inherent paleoclimatic information that these abundance patterns provide, they may also contain some useful stratigraphic information. However, because climatic change can be both global and regional in nature, the stratigraphic usefulness of abundance fluctuations of individual species may be restricted to a limited geographic region. For example, late Pleistocene fluctuations in the abundance of *Neogloboquadrina dutertrei* in the eastern Mediterranean have been associated with major changes in sea-surface salinities and provide an excellent means for stratigraphically correlating deep-sea sequences from throughout this region (Thunell et al. 1977). However, these salinity changes are localized events only, and the N. *dutertrei* stratigraphy therefore can not be extended beyond this region.

The relative abundances of G. *menardii* (warm-water species) and *Globorotalia inflata* (cool-water species) have been used commonly in the Gulf of Mexico for both paleoenvironmental and stratigraphic purposes. These two species are considered to be the most temperature-sensitive planktonic foraminifera in the Gulf (Akers 1965; Beard 1969 and 1973), with their inverse abundance patterns recording the major Pleistocene glacial/ interglacial cycles. In fact, Beard (1973) has used the abundance ratio of these two species to correlate Gulf of Mexico climatic stages with equivalent units from the Caribbean, Atlantic, and sub-Antarctic. The most recent climatically induced reduction in G. *inflata* and increase in G. *menardii* occurs near the Z/Y boundary, at approximately 11 KYBP.

As has been discussed previously, the G. *menardii* zonation of Ericson and Wollin (1968) allows for a tenfold subdivision of the Pleistocene when the criterion is purely the presence or absence of this species (fig. 2.4). Through a quantitative approach, the resolution of this zonation can be enhanced substantially. Specifically, there is potential for further subdivision of the "interglacial episodes" (R, T, V, X, and Z) if the relative abundance of G. *menardii* is taken into consideration.

This type of approach was used by Kennett and Huddleston (1972a) in their quantitative study of late Pleistocene sediments from the Gulf of Mexico. Their subdivision of zones X and Z can be carried out for the most part by simply determining the relative abundance of G. *menardii* (fig. 2.5). Zones X1, X3, X5, and Z2 typically contain higher abundances of G. *menardii* than do X2, X4, and Z1. The resultant zonation is a significant improvement over that based simply on the presence or absence of G. *menardii*. Because this species is a warm-water form, variation in its abundance is a useful tool only for subdividing interglacial episodes. Frequency of variations in the cool-water species, G. *inflata*, may provide a useful equivalent for subdivision of the glacial episodes. The recent study of Neff (1983) on a drilled core from the De-Soto Canyon area clearly demonstrated the inverse relationship in the abundance patterns of G. *menardii* and G. *inflata* through-

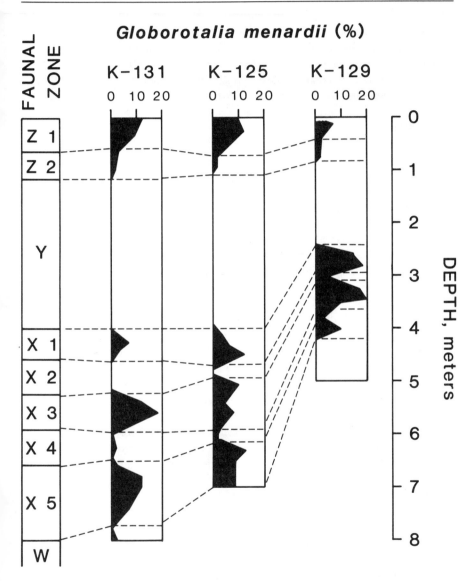

Figure 2.5 *Intercore correlation of abundance variations in the planktonic foraminifera* Globorotalia menardii *for three cores from the Gulf of Mexico (based on Kennett and Huddleston 1972).*

out the Pleistocene (fig. 2.6). In fact, these two species form the basis for a more detailed Pleistocene planktonic foraminiferal zonation recently developed by Neff (1983) for the Gulf of Mexico (see the section on Total Faunal Approach in this chapter for a more complete discussion of this zonation).

Coiling Ratios
Coiling direction is probably the most commonly measured morphologic feature of planktonic foraminifera that is used for both paleoenvironmental and stratigraphic purposes (see Kennett 1976, for a review). Changes in coiling direction have been attributed to phenotypic responses to changing environmental conditions, as well as genetic factors. A stratigraphically useful coiling change occurs in the *G. menardii* plexus in the late Pliocene, with the dominantly dextral population of this group being replaced by sinistral forms. This change in coiling direction has been observed in the Gulf of Mexico (Lamb and Beard 1972; Brunner and Keigwin 1981), as well as the Caribbean and the equatorial Atlantic (Bolli and Premoli-Silva 1973; Parker 1973). As such, this coiling change serves as a reliable stratigraphic horizon for interregional correlation and aids in identifying the general vicinity of the Pliocene/Pleistocene boundary.

For Pleistocene work, coiling changes in three species, *Neogloboquadrina pachyderma*, *G. truncatulinoides* and *P. obliquiloculata* appear to contain the most useful paleoenvironmental and stratigraphic information. Once it was established by Ericson (1959) and Bandy (1960) that dextral assemblages of *N. pachyderma* dominate temperate and subtropical regions, and sinistral populations dominate polar and subpolar areas, coiling changes in this species through time became a very commonly used paleoclimatic index. Ingle (1967 and 1973) has demonstrated that coiling changes in *N. pachyderma* provide a useful means for correlating late Cenozoic sediments from throughout the North Pacific. In a similar fashion, Jenkins (1967) subdivided the late Miocene to Recent of New Zealand into ten zones based on consistent coiling changes in *N. pachyderma*. Owing to its preference for cold water masses, *N. pachyderma* generally is not very abundant in the Gulf of Mexico, with left-coiling forms being particularly rare. As a result, coiling changes in this species are of little value in Pleistocene studies of the Gulf.

Figure 2.6 *Downcore variations in the percent abundance of the planktonic foraminifera* Globorotalia inflata *and* Globorotalia menardii *for drilled core E67–135 from the DeSoto Canyon (based on Neff 1983). The zonation of Ericson and Wollin (1968) is plotted on the right.*

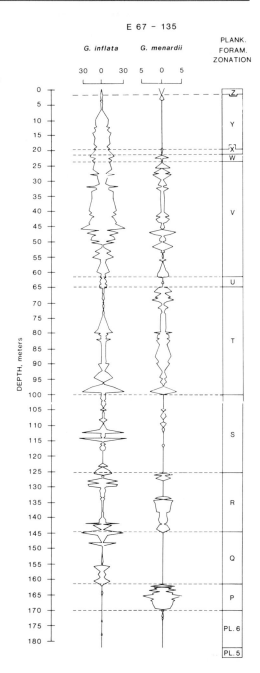

Coiling changes in G. *truncatulinoides* were first shown to be of value in correlating Pleistocene sediments from the Caribbean and the equatorial Atlantic (Ericson and Wollin 1956a; Ericson et al. 1954). This parameter was subsequently used in the Gulf by Ewing and others (1958), Beard (1969), Lamb (1969), Lamb and Beard (1972), Kennett and Huddleston (1972a), Brunner and Keigwin (1981), and Beard and others (1982). Within the Gulf, the right-coiling variant of G. *truncatulinoides* has been the dominant form throughout the Pleistocene, and therefore it is the brief intervals during which left-coiling forms become abundant that are stratigraphically useful. Lamb and Beard (1972) have recognized five major "sinistral events" in G. *truncatulinoides* that are of stratigraphic significance. In a general way, the right-coiling, or dextral, forms are associated with warm water masses, whereas the sinistral forms are more common in cool water masses. Figure 2.7 illustrates an attempt by Kennett and Huddleston (1972a) to use coiling changes in G. *truncatulinoides* to correlate late Pleistocene sequences from the Gulf of Mexico.

Coiling changes in the genus *Pulleniatina* provide another valuable tool for Pleistocene stratigraphy. This faunal parameter has been used most extensively in the Pacific, with five major coiling events occurring between the beginning of the Pleistocene and 785 KYBP (Saito 1976; Thompson and Sciarillo 1978). Since then, this species has been exclusively dextral in this region. In the Gulf region there appear to be fewer coiling changes in *Pulleniatina* during the Pleistocene. In fact, there are only two short intervals, in the latest Pliocene and earliest Pleistocene, when sinistral forms of *Pulleniatina* occur in the Gulf of Mexico (Lamb and Beard 1972; Brunner and Keigwin 1981; and Neff, 1983). However, these events can be found consistently throughout the Gulf and therefore provide additional biostratigraphic means for recognizing the vicinity of the Pliocene/Pleistocene boundary. In addition, these two coiling changes in P. *obliquiloculata* were identified also in the subtropical Atlantic (Saito et al. 1975) and thus provide a reliable means for interregional correlation.

Total Faunal Approach
The relative abundance of an individual species (i.e., G. *menardii*) can be used for stratigraphic purposes as was discussed above. This approach can be elaborated by considering the relative

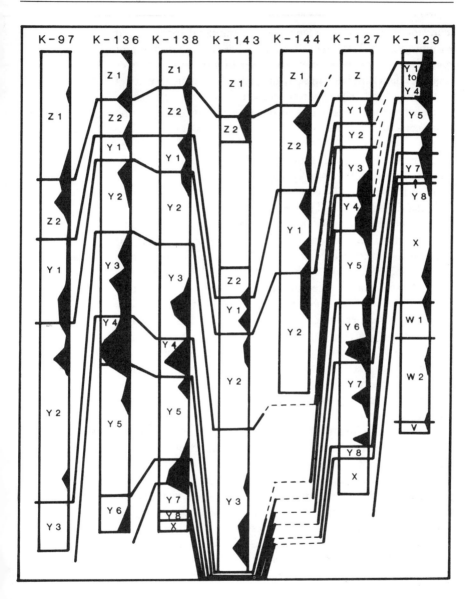

Figure 2.7 *Intercore correlation of changes in the coiling ratio of the planktonic foraminifera* Globorotalia truncatulinoides *for a series of piston cores from the Gulf of Mexico (from Kennett and Huddleston 1972). 100% left-coiling is to the left of each column; 100% right-coiling is to the right of each column. (Reprinted by permission of The University of Washington.)*

abundances of all of the planktonic foraminiferal species in an assemblage (or at least all of the species that typically account for more than several percent of the assemblage). The census or frequency data generated in such an approach contain a wealth of stratigraphic and paleoenvironmental information. It must be remembered that the changes in species abundances during the Pleistocene are primarily a reflection of changing climatic conditions.

For the Gulf of Mexico, the best example of using the abundance patterns of numerous species to create a biostratigraphic zonation is the work of Kennett and Huddleston (1972a). Their study clearly demonstrated that the frequency oscillations of most species are highly correlatable from one section to another. Using the eight most dominant taxa, Kennett and Huddleston (1972a) were able to subdivide the last 195,000 years into 18 faunal zones. Figure 2.8 illustrates in a schematic fashion the species abundance changes that are used in recognizing each of these zones. Although the W–Z zones provide the basic superstructure for this scheme, the Kennett and Huddleston zonation has more than four times the stratigraphic resolution possible with the original Ericson and Wollin zonation, with each subzone having an average duration of less than 15,000 years. In addition to its obvious stratigraphic contribution, the study of Kennett and Huddleston (1972a and b) produced a detailed paleoclimatic history for the late Pleistocene of the Gulf of Mexico (see the section in this chapter on Paleoclimatology).

Following Kennett and Huddleston (1972a), Neff (1983), in recent work, used a total fauna approach to establish a high-resolution biostratigraphic scheme for the Gulf, back to the beginning of the Pleistocene. In this study, zones Q–V were divided further into 15 subzones on the basis of frequency oscillations in seven planktonic foraminiferal species (fig. 2.9). G. inflata and G. menardii were again the two species that were most useful in developing this higher-resolution zonation. The average duration for each of these zones is approximately 100,000 years. Although this does not approach the stratigraphic resolution available through the Kennett and Huddleston zonation for the late Pleistocene, it is a significant improvement over the original Ericson and Wollin zones, which have an average duration of nearly 300,000 years.

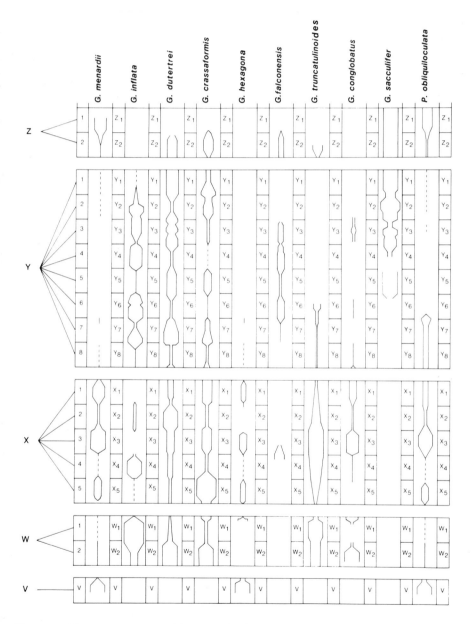

Figure 2.8 *Schematic representation of the species abundance patterns associated with the Kennett and Huddleston (1972) late Pleistocene planktonic foraminiferal zonation for the Gulf of Mexico.*

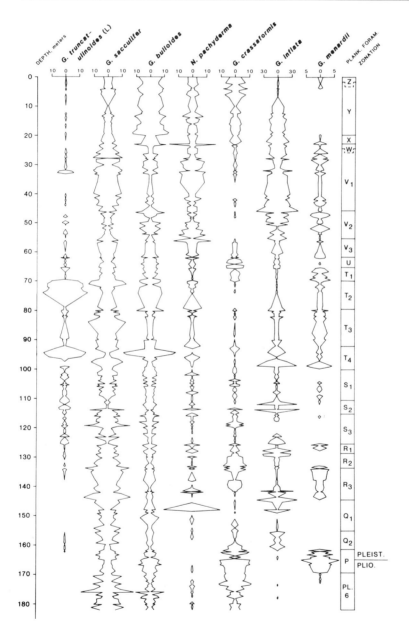

Figure 2.9 *Downcore variations in the percent abundance of 7 planktonic foraminiferal species in E67–135 used to define the Pleistocene planktonic foraminiferal zonation plotted on the left (based on Neff 1983).*

Although most of the zones defined by Neff (1983) are based on multiple species criteria, two of the zones can be identified simply by their anomalously high abundances of sinistral G. *truncatulinoides*. This species comprises 15–20% of the assemblage in zones T2 and T4, in comparison to only several percent (2–5%) of the total assemblage throughout the remainder of the Pleistocene (fig. 2.7). It would appear that these two G. *truncatulinoides* "acme zones" have considerable biostratigraphic potential within the Gulf.

Paleoclimatology

The Pleistocene paleoclimatic history of the Gulf of Mexico is dominated in an obvious way by the periodic oscillation between glacial and interglacial modes. Because it is a marginal sea, the paleoclimatic picture contained in the sedimentary record from the Gulf is a combination both of global climatic events and more localized events that impacted surrounding land masses. As has been discussed previously, Pleistocene planktonic foraminiferal biostratigraphy is based to a large extent on climatically related faunal changes. Thus, Pleistocene planktonic foraminiferal biostratigraphy and paleoclimatology are intimately related. The faunal data necessary to utilize any of the biostratigraphic techniques discussed will yield a certain amount of paleoclimatic information. In the following sections of this chapter, some of the approaches that have been used to reconstruct the Pleistocene paleoclimatic history of the Gulf will be reviewed, and some of the major paleoclimatic trends will be discussed.

Present-Day Planktonic Foraminiferal Biogeography of the Gulf of Mexico

As is the case with all paleoecological studies based on planktonic foraminifera, the ability to make reasonable inferences about past climatic conditions presupposes a thorough knowledge of the modern ecology and biogeography of this group. The present-day biogeographic distributions of planktonic foraminifera in the world's oceans, together with their inferred environmental preferences, has been reviewed by Bé (1977). Within the Gulf of Mexico, more detailed studies of the present-day distribution of planktonic foraminifera and their relationship to hydrographic conditions (i.e., T°C, S ‰) have been carried out by Snyder (1978) and Brunner (1979). Despite the fact that the Gulf as a whole can be considered part of the subtropical province, it does contain faunal elements that have a relatively wide range of temperature preferences. Figure 2.10 illustrates the present-day distributions of some of the more "climatically sensitive" planktonic foraminiferal species in the Gulf of Mexico. These biogeographic patterns are clearly a reflection of the prevailing hydrographic conditions in the Gulf (fig. 2.11). It would appear that winter conditions are primarily responsible for the observed distributional patterns, because summer surface temperatures, and to a lesser extent summer surface salinities, are very uniform throughout the Gulf of Mexico.

When one is dealing with a large number of species it is often difficult to compare distributional patterns objectively and determine what underlying relationships exist among species. Instead of talking in terms of distributional patterns of individual species, it is often more meaningful to be able to identify assemblages of species that are associated with specific water-mass conditions. This can be achieved most easily through some sort of multivariate analysis, such as factor analysis (Imbrie 1963; Imbrie and Kipp 1971; Klovan and Imbrie 1971; Klovan and Miesch 1976) of the species-distributional data to define naturally occurring species assemblages. This approach has been carried out for the Gulf of Mexico by Brunner (1979), and figure 2.12 illustrates the areal distribution of what are interpreted as tropical and temperate assemblages. The temperate assemblage consists of *G. inflata*, *Globigerina bulloides*, *Globigerina falconensis* and *Globerina truncatulinoides*, and is most prominent in the northern Gulf, where winter temperatures are lowest (fig. 2.11). The tropical assemblage, which is dominated by *Globigerinoides sacculifer*, is most abundant in the southern and eastern sectors of the Gulf, where the Loop Current brings in warm surface waters from the Caribbean and the equatorial Atlantic.

This understanding of the relationship between modern biogeographic distributions and surface water conditions provides a basis for making accurate Pleistocene paleoclimatic interpretations from fossil assemblages.

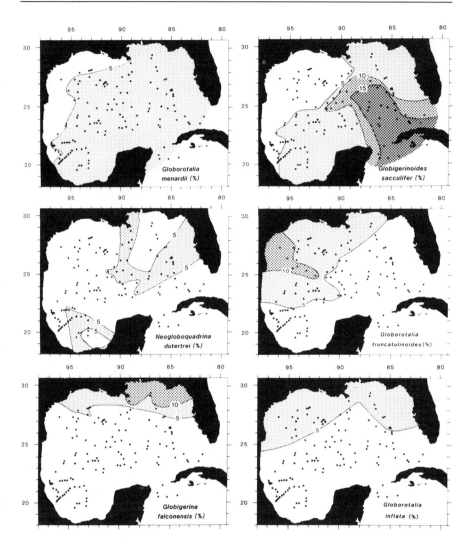

Figure 2.10 *Abundance patterns of six planktonic foraminiferal species in surface sediments from the Gulf of Mexico. (From Snyder 1978 and Brunner 1979. Reprinted with permission of Micropaleontology Press.)*

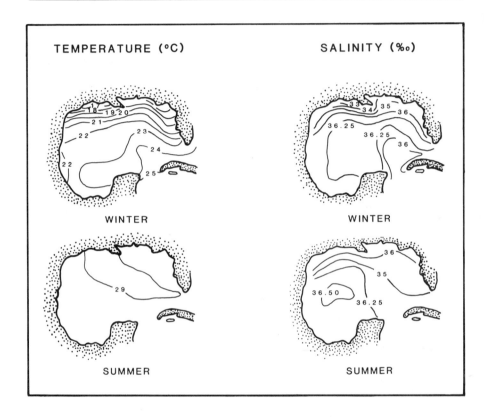

Figure 2.11 *Mean sea surface temperature and salinity patterns for the winter and summer in the Gulf of Mexico (based on data from the National Oceanographic Data Center, 1966).*

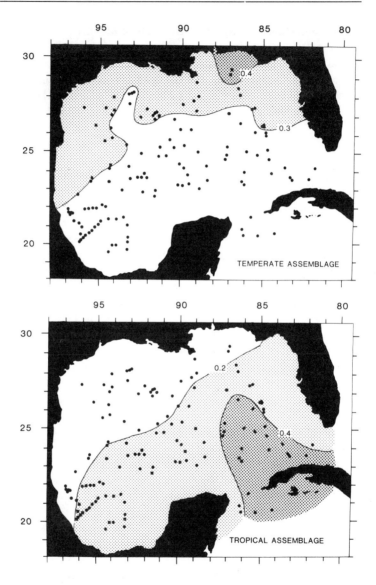

Figure 2.12 *Present-day distribution of temperate and tropical assemblages in Gulf of Mexico surface sediments (from Brunner 1979, reprinted with permission of Micropaleontology Press). The assemblages were derived through a factor analysis of individual species abundance data.*

Pleistocene Paleoclimatology of the Gulf of Mexico

Schott (1935) was really the first to demonstrate that changes in faunal composition were related to changing climatic conditions. In particular, he discovered that G. *menardii* had been intermittently present in the Atlantic during the late Pleistocene and that its presence denoted interglacial, or warm, periods. As already has been discussed, this observation forms the basis for the subsequent Pleistocene planktonic foraminiferal zonation of Ericson and Wollin (1968).

The first systematic studies of late Pleistocene paleoclimates in the Gulf were carried out in the 1950s by Phleger (1951), Parker (1954), and Ewing and others (1958). By utilizing Schott's (1935) findings, these workers were able to identify alternating glacial and interglacial episodes by monitoring changes in the relative abundance of certain "climatically sensitive" planktonic foraminiferal species. Specifically, all of these workers concluded that for the Gulf of Mexico G. *menardii* is the best warm-water indicator and G. *inflata* is the best cold-water indicator. This relationship was used in many subsequent studies of Pleistocene paleoclimates in the Gulf (Beard 1969 and 1973; Kennett and Huddleston 1972a and b; Neff 1983). Figure 2.6 clearly illustrates the inverse relationship that exists in the relative abundances of these two species throughout the Pleistocene.

This type of individual-species approach provides a relatively rapid and reliable way of identifying major climatic trends. A variety of other techniques, which generally utilize frequency data for all or most of the species in an assemblage, are available for resolving the more detailed features of the paleoclimatic or paleotemperature record. The simplest of these approaches is to assign species to either a warm or a cold faunal group on the basis of their modern biogeography, and then to calculate the frequency ratio of these groups. This approach was used first by Ruddiman (1971) in the Atlantic, and subsequently by Beard (1973) and Beard and others (1982) in the Gulf of Mexico to produce paleotemperature records for the entire Pleistocene that have been correlated with the North American glacial stages. This method, however, weights all species as being equally "warm" or "cold," and thus shows the disadvantage of not being able to recognize subtle differences between these two end members. Kennett and Huddleston (1972a) attempted to resolve this problem by dividing the Pleistocene planktonic foraminifera of the Gulf into five categories on the basis of the temperature preferences of the individual species. The paleoclimatic record for the last 200,000 years produced using the relative frequencies of all of these temperature-related groups correlates very well the Caribbean oxygen isotope record of Emiliani (fig. 2.13). This record provides a detailed picture not only of the timing of late Pleistocene paleoclimatic changes in the Gulf, but also of the relative magnitude of these changes.

Another way of evaluating the abundance data for numerous species is with some sort of multivariate statistical analysis. The goal of this approach is really much the same as that produced by simply grouping environmentally similar species together. The major difference is that the grouping of species is done "objectively" by the multivariate analysis, whereas it may be somewhat "subjective" in the basic total faunal approach. Factor analysis, which was referred to earlier, and principal component analysis are the two multivariate techniques that are most commonly used on micropaleontologic census data (Klovan and Imbrie 1971; Davis 1973). The major objective of both of these analyses is to reduce a large number of variables, in this case planktonic foraminiferal species, into a smaller number of independent factors or components that identify major interrelationships among the different species through time. The species groupings determined by these analyses will, it is hoped, reflect naturally occurring assemblages that can be interpreted on the basis of our knowledge of their present-day ecology. Geometrically, principal components or factors can be thought of as vectors corresponding to the primary axes of a multidimensional data set. Thus, the first factor or component is the vector that passes through the direction of maximum variance for a given data set; essentially a best-fit straight line. Because the factors or components are ranked according to the amount of variance they explain, and temperature is responsible for most of the variability in species abundances during the Pleistocene, the first factor or component will generally yield a paleotemperature record. These statistically derived paleoclimatic records are often quite similar to those produced by using other faunal and isotope techniques (fig. 2.13).

The principal component technique has been used by Thunell (1976) and Malmgren and Kennett (1976) on planktonic

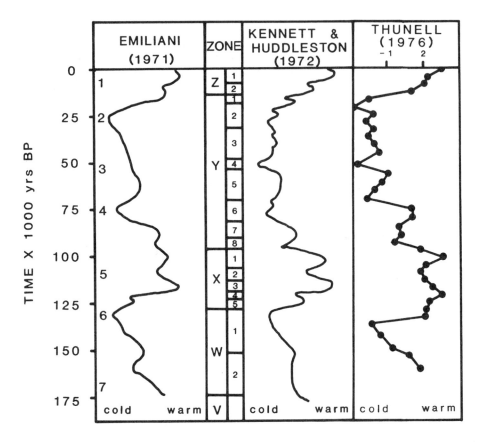

Figure 2.13 *Comparison of late Pleistocene paleoclimatic curves. The Emiliani (1971b) curve is an oxygen isotope record for a Caribbean piston core. The Kennett and Huddleston (1972) curve is based on the relative abundance of planktonic foraminiferal species in Gulf of Mexico piston cores. The Thunell (1976) curve is based on a principal component analysis of species abundance data from a Gulf of Mexico piston core. In all cases, warm is to the right and cold is to the left. The zonation is from Kennett and Huddleston (1972). The time scale is from Broecker and Van Donk (1970).*

foraminiferal abundance data from Pleistocene age Gulf cores. In both studies, planktonic foraminiferal assemblages were derived that are clearly temperature related. The "warm-water" assemblage contained G. *menardii*, P. *obliquiloculata*, G. *truncatulinoides* and N. *dutertrei*, whereas the "cool-water" assemblage was made up of G. *inflata*, G. *falconensis*, G. *bulloides*, *Hastigerina aequilateralis* and *Orbulina universa*. Figure 2.14 illustrates the first principal component or paleoclimatic record for four late Pleistocene cores from the Gulf. When sampling resolution is adequate, much of the detail in these climate records can be correlated between cores. For example, the three warm intervals and two cool intervals associated with isotope stage 5 (see chap. 3) are readily observable in these faunal paleoclimatic records (fig. 2.14).

The concept of using statistical techniques to identify temperature-related planktonic foraminiferal assemblages can be carried one step further, to estimate absolute paleotemperatures. Modern planktonic assemblages defined by factor analysis can be mathematically regressed against the temperatures of the surface water in which they lived, to produce temperature transfer-function equations. These equations can then be used to calculate paleotemperatures from fossil planktonic foraminiferal assemblages. The underlying assumption in this approach is that an individual species temperature preference has remained relatively unchanged throughout the Pleistocene. This approach was used by CLIMAP (1976 and 1981) to reconstruct surface temperatures of the world's ocean during the last glacial maximum. Within the Gulf of Mexico, Brunner and Cooley (1976) and Brunner (1979) have derived transfer functions for estimating late Pleistocene surface-water temperature fluctuations. Brunner (1982) has demonstrated that during the last glacial maximum (18 KYBP), temperatures in the Gulf were 1°C to 2°C cooler than at present. In contrast, during the last climatic optimum (125 KYBP) surface temperature patterns were the same as those of the present day.

Much of the paleoclimatic work that has been done in the Gulf of Mexico is based on piston core material and therefore is restricted in time to the late Pleistocene. As a result, our understanding of pre–late Pleistocene climates in this area is rather limited. Beard (1969) and Beard and others (1982) have used planktonic foraminifera to define warm-cold cycles for the Pleis-

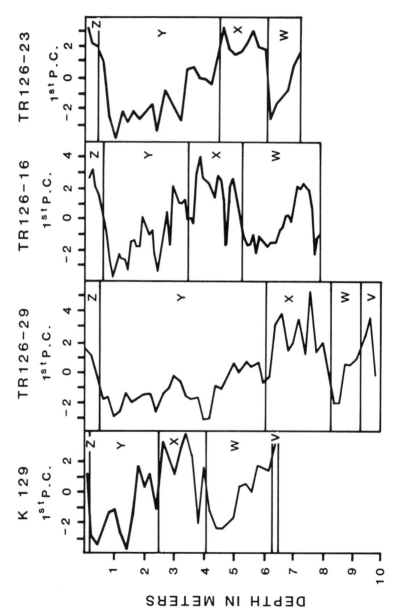

Figure 2.14 Relative paleotemperature curves for four Gulf of Mexico piston cores generated from a principal component analysis of planktonic foraminiferal species abundance data (from Malmgren and Kennett 1976). In all cases warm is to the right and cold is to the left. The positions of the Globorotalia menardii zones of Ericson and Wollin (1968) are indicated for each one. (Reprinted by permission of Elsevier Scientific Publishing Company.)

tocene that in turn have been correlated to eustatic sea-level changes and the North American glacial stages (fig. 2.15). Unfortunately, the resolution of these studies is rather limited, and as a result, they provide us only a very general picture of changing climatic conditions in the Gulf during the Pleistocene.

The recent work of Neff (1983) represents the first high-resolution planktonic foraminiferal paleoclimatic study for the entire Pleistocene epoch in the Gulf region. This study clearly indicates that the individual zones within the Ericson and Wollin zonation do not represent strictly glacial or strictly interglacial conditions, but each zone rather incorporates both glacial and interglacial climates. Of particular significance is Neff's (1983) observation that the Upper Pleistocene was cooler and marked by greater climatic variability than was the Lower Pleistocene. This is seen most simply as an overall increase in the abundance of the cool-water species *G. inflata* in the Upper Pleistocene (fig. 2.16). A paleoclimatic record generated from a factor analysis of the species-abundance data from a core drilled in the DeSoto Canyon also reveals this same climatic pattern (fig. 2.16). This shift in climatic regimes between the Lower and Upper Pleistocene is not peculiar to the Gulf of Mexico but also has been observed in the Pacific (Shackleton and Opdyke 1976; Prell 1982); the Atlantic (Briskin and Berggren 1975; van Donk 1976; Shackleton and Cita 1979; Prell 1983); and the Mediterranean (Thunell and Williams 1983). It has been proposed that the establishment of perennial sea-ice cover in the Arctic, combined with large continental ice sheets, is responsible for this Middle-Pleistocene climatic shift and the intensification of glacial/interglacial cycles (Herman and Hopkins 1980; Williams and others 1981).

Summary

Since the early 1950s, planktonic foraminifera have been used extensively in both biostratigraphic and paleoclimatic studies of the Pleistocene of the Gulf of Mexico. During this time period, a gradual shift away from a qualitative approach to micropaleontology occurred in favor of more quantitative techniques, capable of providing greater biostratigraphic and paleoclimatic resolution.

Despite the confusion that has existed over the years in the

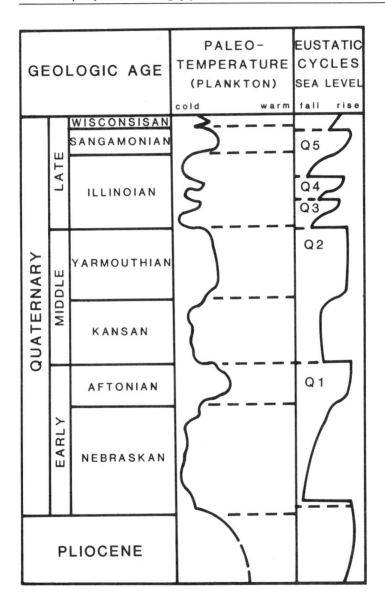

Figure 2.15 *Comparison of North American Pleistocene glacial stages, planktonic foraminiferal paleotemperature changes and eustatic sea-level cycles (based on Beard et al. 1982).*

Figure 2.16 *Comparison of*
Globorotalia inflata *and*
Globorotalia menardii *species abun-*
dance patterns with a paleoclimatic
record derived from a factor analysis
of all the downcore species abundance
data for E67–135. For the paleo-
climatic record, colder conditions are
indicated by higher values for Factor
II. The Globorotalia menardii *zona-*
tion of Ericson and Wollin (1968) is
indicated.

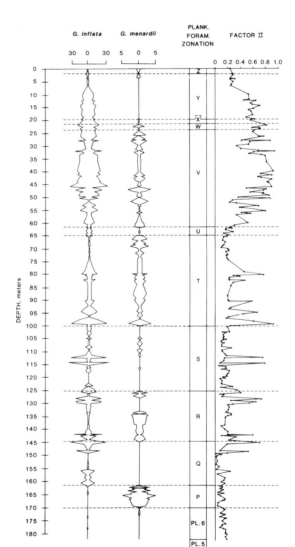

recognition of the Pliocene/Pleistocene boundary in the Gulf of Mexico, there are a number of reliable planktonic foraminiferal datums (last appearance of G. *obliquus* and the first appearance of G. *truncatulinoides*) that can be used to identify the boundary in the Gulf and to correlate the Gulf record with that of the open ocean. Calibration of these datums to the polarity time-scale indicates that the age of the Pliocene/Pleistocene boundary is approximately 1.6 to 1.8 MYBP, and not associated with the development of major Northern Hemisphere glaciation, which began in the late Pleistocene.

The Ericson and Wollin (1968) zonation, which is based on the presence-absence pattern of G. *menardii* and allows for a tenfold subdivision of the Pleistocene, has served as the primary biostratigraphic framework for the Gulf Pleistocene. The more recent work of Kennett and Huddleston (1972) and Neff (1983) has demonstrated that a quantitative approach can improve the resolution of the original Ericson and Wollin zonation substantially. Specifically, by evaluating the frequency patterns of the more abundant planktonic foraminifera, Kennett and Huddleston (1972) were able to subdivide zones W–Z (approximately, the last 195,000 years) into a total of 18 subzones. Likewise, Neff (1983) was able to subdivide zones Q–V of Ericson and Wollin (1968) into 15 faunal units, with the newly defined zones having an average duration of approximately 100,000 years, in comparison to 300,000 years for the original zones.

3

Correlation of Pleistocene Marine Sediments of the Gulf of Mexico and Other Basins Using Oxygen Isotope Stratigraphy

Douglas F. Williams
Department of Geology and Marine Science Program
University of South Carolina
Columbia, South Carolina 29208

3

Introduction

Over the last two decades, the extensive study of oxygen isotope records in numerous deep-sea cores has led to the development of a precise stratigraphic method capable of establishing global and regional correlations of Pleistocene marine sediments. Precise correlations can now be made within an absolute time framework with a resolution of 10,000 to 20,000 years during the late Pleistocene (the last 1 million years) and of less than 40,000 to 50,000 years during the early Pleistocene—between 1.0 and 1.8 million years before present (MYBP). Establishment of this important stratigraphic tool is possible owing to (1) the availability of detailed oxygen isotope records from each of the major ocean basins as well as from marginal seas like the Gulf of Mexico, the Mediterranean, and the Labrador Sea and (2) a better understanding of the factors that determine the oxygen isotope signal locked in the remains of microfossils commonly found in marine sediments. Oxygen isotope records are most often based on stable isotopic analyses of fossil benthic and planktonic foraminifera, minute protozoans that form calcium carbonate shells or tests. Reliable records also have been obtained from isotopic analyses of fossil calcareous nannoplankton (minute phytoplankton that secret calcareous plates known as coccoliths) and fossil diatoms (phytoplankton that form delicate tests made of opaline silica).

Besides being used in stratigraphic correlation and absolute-age dating, oxygen isotope records also provide direct evidence of glacial-interglacial paleoclimatic cycles and glacio-eustatic sea-level changes. For example, figure 3.1 shows an example of how the oxygen isotope record becomes more positive during glacial times, when increased amounts of water are stored as ice volume in high-latitude areas during global lowering of sea level. At the termination of glacial conditions, the $^{18}O/^{16}O$ ratio rapidly becomes less positive as water stored as ice in high-latitude regions returns to the oceans and sea level rapidly rises to interglacial levels (fig. 3.1). These cycles are repeated throughout the Pleistocene and are well dated within paleomagnetic and biostrati-

This chapter appears by permission of the Colorado School of Mines, Golden, Colorado. A version of this will be published in a forthcoming work, *Subsurface Geology*, edited by LeRoy and LeRoy, 5th edition.

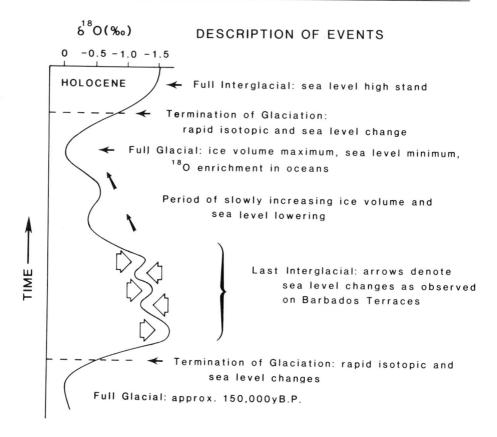

Figure 3.1 *Schematic representation of the oxygen isotope record for the last 150,000 years and its approximate relationship with major climatic and sea-level changes of the late Quaternary.*

graphic frameworks (see chaps. 1 and 2). The oxygen isotope record thereby provides a precise record of the transgression-regression cycles that typified the Pleistocene. Changes in sedimentation patterns and the lithologic character of shelf and continental-margin sediments thus can be interpreted within the glacio-eustatic history of an ocean basin.

The major objectives of this chapter, therefore, are to provide an up-to-date synthesis of the Pleistocene oxygen isotope record and to illustrate the use of oxygen isotope stratigraphy for correlation and paleoenvironmental interpretation, especially with regard to the Gulf of Mexico. First, basic information will be provided concerning the generation of oxygen isotope data from foraminifera to introduce some of the terminology used in oxygen isotope stratigraphy. Next, the historical development of oxygen isotope stratigraphy will be described, to clearly distinguish it from the original paleotemperature work. Finally, oxygen isotope records from various regions and time intervals of the Pleistocene will be described, to illustrate the interpretations and stratigraphic usefulness of the method. Although much basic work remains to be done in Cenozoic and Mesozoic marine sections, the isotopic technique shows potential application in stratigraphic studies of older marine sequences. Interesting information may also be gained from the carbon isotope ratio ($^{13}C/^{12}C$) of calcareous microfossils, but this aspect is beyond the scope of this review. I have attempted to make this review as complete as possible by extensive references throughout the text, but see the excellent review articles by Hecht (1976), Duplessy (1978), Berger (1979) and Savin and Yeh (1981), each of which contains supplementary information not possible to cover in this chapter.

Methodology

Generation of the Oxygen Isotope Data
The basic analytical technique for determining the $^{18}O/^{16}O$ ratio of calcium carbonate ($CaCO_3$) was developed in 1950 by McCrea and has been modified over the last 20 years until the technique is now fairly uniformily established in stable isotope laboratories doing oxygen isotope analyses of foraminifera. Usually, 5–40 individual tests of foraminifera (approx. 0.3–1 mg $CaCO_3$ by

weight) are needed for an isotopic analysis, depending on the inlet system of the available mass spectrometer. The foraminifera should be cleaned of fine-grained sediment by gentle use of an ultrasonic device, either in methanol (Shackleton and Opdyke 1973) or distilled water. This procedure is particularly important in calcareous oozes in which substantial quantities of calcareous nannoplankton may be present in the foraminiferal tests. The foraminifera are then air-dried or dried in an oven at temperatures less than 50°C. Each sample is then roasted under vacuum at temperatures approaching 380°C, to remove any organic material that may be associated with the fossil tests. The carbonate is then reacted in an evacuated reaction vessel with purified phosphoric acid (H_3PO_4) held at a constant temperature (usually 50°C or 25°C). Reagent grade 85% H_3PO_4 may be purified by heating the acid under vacuum to drive off absorbed water. The reaction of $CaCO_3$ with H_3PO_4 yields carbon dioxide gas and water vapor by the following equation:

$$3CaCO_3 + 2H_3(PO_4) \rightleftharpoons 3CO_2 + 3H_2O + Ca_3(PO_4)_2 \tag{3.1}$$

The resultant carbon dioxide gas must be purified of water vapor by fractional freezing. A series of isopropyl alcohol–liquid nitrogen (or IPA–dry ice) traps are used to freeze the water vapor, while liquid nitrogen is used to transfer the purified CO_2 either directly into an isotope-ratio mass spectrometer or into an ampule of varying design for later isotopic analysis (fig. 3.2).

In the mass spectrometer, the CO_2 molecules are ionized into positively charged ions and separated into ion beams of three different masses (mass 44 = $^{12}C\ ^{16}O\ ^{16}O$; mass 45 = $^{13}C\ ^{16}O\ ^{16}O$; mass 46 = $^{12}C\ ^{16}O\ ^{18}O$). The ratio of the intensities of the mass 46 to mass 44 ion beams in the mass spectrometer yields the $^{18}O/^{16}O$ ratio of the CO_2 gas. The $^{18}O/^{16}O$ ratio of the sample CO_2 is then repeatedly compared to the same ratio in a laboratory reference CO_2 gas under identical operating conditions. By convention, the ratio of the sample CO_2 is expressed as a delta value (δ), or the parts per thousand (per mil, ‰) enrichment or depletion in ^{18}O relative to the reference CO_2:

$$\delta^{18}O\ ‰ = \left[\frac{^{18}O/^{16}O_{sample}}{^{18}O/^{16}O_{reference}} - 1 \right] 10^3 \tag{3.2}$$

CaCO₃ EXTRACTION LINE

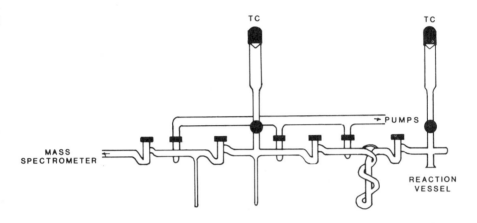

Figure 3.2 *An example of carbonate extraction line for the routine purification of carbon dioxide gas derived from calcium carbonate prior to determination of the stable oxygen and carbon isotope ratios ($^{18}O/^{16}O$ and $^{13}C/^{12}C$). This carbonate line is one of several configurations that may be attached directly to a mass spectrometer (on-line) or may be remote from the mass spectrometer.*

All delta values are described relative to the heavier isotope in terms of "positive," "enriched," or "heavy" (that is, containing more ^{18}O than CO_2 from the reference CO_2. Mass spectrometers available today can determine $\delta^{18}O$ values for CO_2 gas produced by the technique described here with a precision approaching 0.05 ‰ or to much better than 5–10% of the total isotopic change observed in foraminifera from Pleistocene sediments.

To ensure that $\delta^{18}O$ values for foraminifera can be compared between laboratories using different reference gases, every major isotope laboratory corrects the measured delta values for machine effects (Craig 1957; Deines 1977) and calibrates its reference CO_2 gas to the universal PDB standard. All $\delta^{18}O$ values for foraminifera are reported relative to PDB, which was a powdered belemite from the Cretaceous Pee Dee Formation of South Carolina (Urey et al. 1951; Epstein et al. 1953a, b). To guarantee the integrity of the isotope data, each laboratory routinely analyzes a reference carbonate powder, usually on a daily basis, before analysis of each set of foraminifera with unknown oxygen isotope compositions.

Historical Development of Oxygen Isotope Stratigraphy

On the basis of the thermodynamic properties of the three isotopes of oxygen (^{16}O, ^{17}O, ^{18}O) Urey (1947) was the first to propose that the $^{18}O/^{16}O$ ratio of natural substances, like calcium carbonate, when formed in isotopic equilibrium with seawater, could be used as a "geothermometer." Urey, his students, and colleagues, then proceeded to develop empirical paleotemperature equations relating the $^{18}O/^{16}O$ ratio of carbonate from mollusks to the temperature and isotopic composition of the waters in which they secreted their carbonate (McCrea 1950; Urey et al. 1951; and Epstein et al. 1953a, b).

Subsequently, Emiliani (1958) pioneered the application of the isotopic paleotemperature technique to fossil foraminifera from deep-sea sediment cores (fig. 3.3). Each of these records showed times when the isotopic temperatures of the oceans were apparently 5–7 degrees colder than at present, and other times

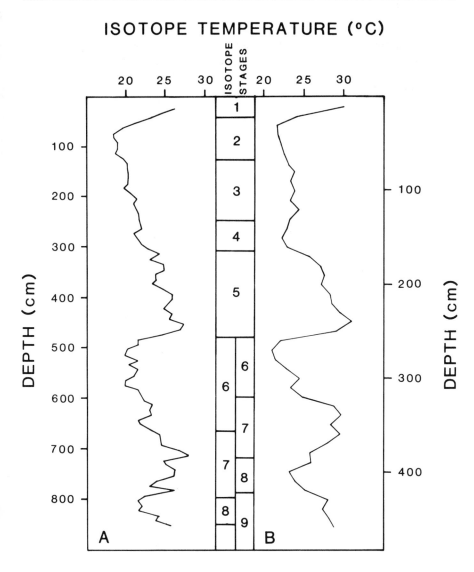

Figure 3.3 *An example of two late Quaternary oxygen isotope records interpreted by Emiliani (1958) in terms of paleotemperatures. The numbers in the center column illustrate the concept of oxygen isotope stages wherein interglacial climatic conditions are represented by odd numbers, and glacial-like climatic conditions are represented by even numbers. The oxygen isotope stages can be recognized in the cores in spite of a greater than twofold difference in sedimentation rate between the cores. Cores A and B (core 246 and A179–4) are from the equatorial Atlantic and the Caribbean Sea, respectively (Emiliani 1958).*

when they were similar to today's interglacial temperatures. His work, more than that of any other single investigator, provided the isotope data of numerous cores and different foraminiferal species, to show that these isotope changes were worldwide (Emiliani 1955, 1956, 1958, 1961, 1964, 1966, 1971, 1972, among others). It is also important to realize that the bulk of this work was performed before the development of the so-called micromass isotope-ratio mass spectrometers available today. His work was done at a time when tremendous quantities of clean, monospecific foraminiferal tests were required for a single isotopic analysis. From these isotopic records, Emiliani recognized and established the concept of "isotopic stages." Times of isotopic temperatures similar to those of today were designated as *interglacials* and given odd numbers; *glacial* stages were even-numbered (fig. 3.3). These isotopic stages later formed the basis for oxygen isotope stratigraphy, because they could be recognized in sediments of every ocean basin, despite differences in sediment accumulation rate.

What began as the largest limitation to the isotopic paleotemperature method, namely uncertainties in estimating the isotopic composition of seawater in past oceans, became the impetus for oxygen isotope stratigraphy. Emiliani estimated that approximately one-third (0.4–0.5 ‰) of the Pleistocene glacial-interglacial isotopic change recorded in deep-sea cores (1.5–2.0 ‰) was due to the removal of water from the oceans during glacial periods and storage of that water in continental ice sheets. The remaining two-thirds of the signal, approximately 1.4–1.6 ‰, was attributed by Emiliani to temperature changes of the ocean waters. Using these values and the ^{18}O-temperature fractionation of 0.22 ‰ per degree Celsius paleotemperature decreases as large as 8°C were postulated during glacial times for equatorial and low-latitude regions of the oceans (fig. 3.3) (Emiliani 1955; 1958; 1966; and other work).

Other investigators argued, however, that the isotopic record of foraminifera primarily reflects changes in the $^{18}O/^{16}O$ composition of seawater as a function of global ice volume, and only secondarily records water temperature changes (Craig 1965; Olausson 1965; Shackleton 1967; Dansgaard and Tauber 1969; Imbrie et al., 1973). The magnitude of this ice effect in the Pleistocene was estimated as 1.2–1.8 ‰ (table 3.1).

Source	Basis	$\delta^{18}O$ ‰
Emiliani, 1955–1974	Ice Sheet Reconstruction and Sea Level	0.4–0.9
Craig, 1965	Ice Sheet Reconstruction	1.5
Olausson, 1965	Ice Sheet Reconstruction	1.7
Shackleton, 1967	$\delta^{18}O$ of Benthic Foraminifera and Ice Sheet Dynamics	1.4
Lidz et al., 1968	$\delta^{18}O$ of Deep Planktonic Species	0.9
Dansgaard and Tuber, 1969	Ice Sheet Reconstruction	1.2
van Donk, 1970	$\delta^{18}O$ of Benthic Foraminifera	0.8–1.2
Imbrie et al., 1973	Faunal Paleotemperature Estimate	1.4
Savin and Stehli, 1974	$\delta^{18}O$ of High-Latitude Foraminifera	0.85
Duplessy, 1978	$\delta^{18}O$ of Benthic Foraminifera	1.1–1.65
Shackleton, 1974, 1977	$\delta^{18}O$ of Benthic Foraminifera	1.65–1.8

Table 3.1 *Estimated Magnitude of the Ice Volume Effect in the Late Pleistocene*

Despite the uncertainties in determining the exact paleotemperature and the effect of ice volume, it is currently believed that 70–90% of the isotopic signal recorded in Pleistocene deep-sea sediments is due to changes in the isotopic composition of seawater as a function of the waxing and waning of continental ice sheets (namely, compositional changes in seawater). Depending on the geographic region, the other 10–30% of the isotope signal may be due to water-temperature change or local variations in the evaporation-precipitation balance of marginal basins.

In addition to the ice effect, Emiliani's chronology also came under criticism, especially the age for the height of the last interglacial (equivalent to isotope stage 5e) (fig. 3.4). Initial age-dating of oxygen isotope records using extrapolated radiocarbon dates (^{14}C) and the radioactive decay of thorium-230 and protactinium-231 indicated that the last interglacial (stage 5e, when climatic conditions were as warm or warmer than those of today) occurred 100,000 years before present (YBP) (Roshalt et al. 1961; Rona and Emiliani 1969; Emiliani and Rona 1969). This chronology was questioned, however, because (1) at that time, radiocarbon dates were reliable only back as far as 30 KYBP, and (2) the ^{230}Th/^{231}Pa–derived sedimentation rates were made by a "leaching" method as opposed to a "total digestion" method (Ku 1966; Broecker and Ku 1969). It was argued by Broecker and his co-workers that erroneously high sediment accumulation rates and therefore, anomalously young age estimates for the last interglacial period, were obtained if the ^{231}Pa/^{230}Th ratio was not corrected properly for the uranium content in each sample so analyzed. An older chronology, based on the method developed by Ku (1966), therefore was used to place the height of the last interglacial (isotope stage 5e) at approximately 120 KYBP, i.e., at least 20,000 years older than the estimate made by Emiliani and co-workers (fig. 3.4).

This older age-estimate for the last interglacial (isotope stage 5e) had great paleoclimatic significance in testing theories of climatic change, and received widespread support from other continuous deep-sea cores with radiometrically determined sedimentation rates (Ericson et al. 1961; Ericson and Wollin 1968; Ku 1966; Broecker 1966; Ku and Broecker 1966; Broecker et al. 1968; Broecker and Ku 1969; Mesolella et al. 1969; Broecker and Van Donk 1970; among others). Radiometric dating of

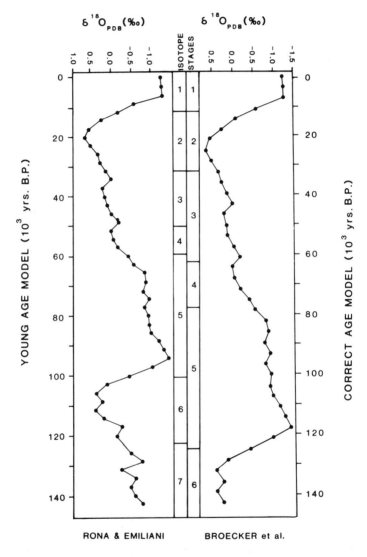

Figure 3.4 *A representation of the early changes in the chronology of the oxygen isotope record for the last 140,000 years using Caribbean core P6304–8 as an example. The age model on the left (A) (Rona and Emiliani 1969) results in ages for the isotope stages which are too young, because of improper corrections to the* ^{231}*Pa/* ^{230}*Th ratios in each sample. The age model on the right (B) for the oxygen isotope stages 1 through 5 is considered to be more correct and is based on a sedimentation rate determined by correcting the* ^{231}*Pa/* ^{230}*Th ratio (Broecker et al. 1969, adapted from Broecker and Ku 1969).*

coral terraces also supported the older date for isotope stage 5e by showing a close relationship between glacio-eustatic sea levels and the oxygen isotope signal in deep-sea sediments (fig. 3.5) (Broecker et al. 1968; Mesolella et al. 1969; Broecker and van Donk 1970; Veeh and Chappell 1970; Bloom et al. 1974; Shackleton and Matthews 1977; Fairbanks and Matthews 1979; among others).

One paper, however, seemed to provide the *central* focus for oxygen isotope stratigraphy as a powerful correlation tool in Pleistocene sediments. In their premier paper, Shackleton and Opdyke (1973) used magnetic stratigraphy (see chap. 1, this volume,) to firmly place the oxygen isotope stages observed in a piston core from the Pacific (V28–238) into an absolute time frame through the past 870,000 years (fig. 3.6). The boundaries of 22 isotope stages were thus dated (table 3.2) and interpreted in terms of large and small volumes of Northern Hemisphere ice. They showed that the isotope data could be used to estimate the magnitude of sea-level changes as well as variations in the volume of ice in high-latitude regions. Since that seminal paper, oxygen isotope records have become among the most sought-after criteria for establishing (1) global correlations of Pleistocene marine sediments (Shackleton 1977; Duplessy 1978; CLIMAP 1976; Savin 1977; Prell et al. 1980; among others); (2) the synchroneity of various micropaleontological events in terms of isochronous last or first appearance datums (Hays and Shackleton 1976; Thierstein et al. 1977; Burckle et al. 1979; Morley et al. 1981; among others; (see also chap. 2, this volume); and (3) the events accompanying the climatic evolution of the late Cenozoic (Savin et al. 1975; Shackleton and Kennett 1975; Savin 1977; Keigwin 1981; Thunell and Williams 1983; among others).

The oxygen isotope record and chronology of V28–238 thus quickly became a kind of "standard" by which all other isotope records were compared, even though the average sediment-accumulation rate in the core was low ($2cm/10^3$ year), and intervals of the core had been affected by calcium carbonate dissolution (Shackleton and Opdyke 1973). Still, the isotope record of V28–238 was the longest and most continuous record available at the time, and efforts were initiated to revise and strengthen the chronology of the isotope stages using the close correspondence between the spectral character of the oxygen isotope record and

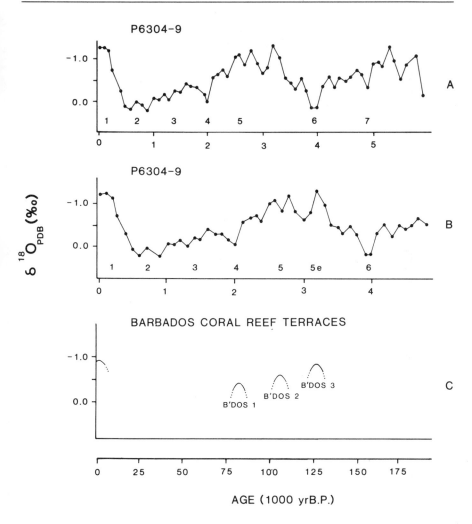

Figure 3.5 An attempt to reconcile the chronological differences in the early development of oxygen isotope stratigraphy with the well-dated late Pleistocene coral reef terraces of Barbados (adapted from Shackleton and Matthews 1977). Plot A represents the young chronology for Caribbean core P6304–9, whereby the warmest part of the last interglacial, isotope stage 5, is considered to be 100 KYBP (Emiliani and Rona 1969; Rona and Emiliani 1969). Plot B shows the older age model for the same oxygen isotope record of P6304–9 whereby isotope stage 5e is more correctly near 127 KYBP. This older chronology confirms the correlation that exists between features of the oxygen isotope record and the high sea-level terraces of Barbados I, II, and III (Part C, from Broecker et al. 1968; Mesolella et al. 1969).

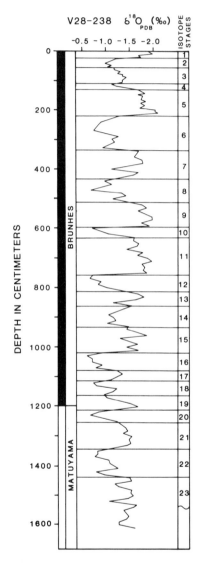

Figure 3.6 *The detailed oxygen isotope record for Pacific core V28–238 illustrating the pioneering attempt by Shackleton and Opdyke (1973) to determine the chronology of the oxygen isotope stages 1 through 23 for approximately the last 1 million years (Table 3.1). An age of 0.69 million years before present was used for the Brunhes/Matayuma paleomagnetic polarity boundary (The age of the boundary was later revised to 0.73 MYBP by Mankinen and Dalrymple 1979.) (Ledbetter, chapter 1. The establishment of this chronology and the recognition of these oxygen isotope stages in other cores led to the concept of oxygen isotope stratigraphy for global stratigraphic correlations (adapted from Shackleton and Opdyke 1973).*

the mathematically determined frequencies in the orbital parameters of the Earth (Hays et al. 1976). These orbital parameters, the so-called Milankovitch parameters, involving the eccentricity, obliquity, and precession of the Earth's orbit, show long-term cycles with frequencies of about 100,000; 40,000; and 23,000 years, respectively. Spectral analysis of the oxygen isotope records for V28–238 (Kominz et al. 1979) and other deep sea cores (Hays et al. 1976) revealed frequencies similar to the orbital frequencies and greatly strengthened earlier work that hypothesized that the Milankovitch parameters played a strong controlling part in causing the rapid climatic changes of the Pleistocene (Mesolella et al. 1969; Broecker et al. 1968; Broecker and Van Donk 1970; among others).

From these important findings, Hays et al. (1976) then ingeniously attempted to use the predictable orbital frequencies to "tune" the chronology of the isotope record (table 3.2). Kominz et al. (1979) also employed the same approach to propose a new chronology for the isotope stages (table 3.2), one based on the isotope record of V28–238 and a revised age date of 728 KYBP for the Brunhes/Matuyama boundary. The most current chronology comes from the work of the SPECMAP Group (Imbrie et al., in press) in which several continuous isotope records were statistically "stacked" into a composite record and tuned to the frequencies characterized by the Earth's orbital parameters. The resulting SPECMAP chronology for the Brunhes Chron shown in table 3.2 will, most likely, become the accepted chronological framework for the late Pleistocene isotope record, barring any unforeseen advances in absolute dating.

Premises of Oxygen Isotope Stratigraphy in Pleistocene Sediments

This section focuses on the properties of the Pleistocene oxygen isotope signal and some of the basic premises of oxygen isotope stratigraphy as summarized below.

1. Past oxygen isotope changes in Pleistocene oceans were driven primarily by glacio-eustatic changes in sea level. As water enriched in the light isotope of oxygen (^{16}O) is preferentially removed from the oceanic reservoir and stored as ice in the polar

Isotope Stage Boundary	Constant Sedimentation[a]	Spectral Analysis[b]	Spectral Analysis[c]	Spectral Analysis[d]
1/2	13	10	11	12
2/3	32	29	29	24
3/4	64	61	61	59
4/5	75	73	73	71
5/6	128	127	127	128
6/7	195	190	190	186
7/8	251	247	247	245
8/9	303	303	303	303
9/10	347	336	336	339
10/11	367	356	352	362
11/12	440	425?	453	423
12/13	472	457?	480	478
13/14	502		500	524
14/15	542		551	565
15/16	592		619	620
16/17	627		649	659
17/18	647		662	689
18/19	688		712	726
B/M	700		728	730
19/20	–		–	736
20/21	–		–	763
21/22	–		–	790

Table 3.2 *Revised Ages of $\delta^{18}O$ Stage Boundaries Based on Time Scales from Various Sources (10^3 years)*

[a]Shackleton and Opdyke, 1973.
[b]Hays et al., 1976.
[c]Kominz et al., 1979.
[d]SPECMAP; Imbrie et al., in press.

regions during glacial periods, the ^{18}O content of the remaining seawater increases and sea level becomes lower (figs. 3.1 and 3.7). The magnitude of the compositional changes in the glacial oceans is proportional to the extent of sea-level lowering and ice volume accumulation (fig. 3.8). The glacio-eustatic sea-level changes are in turn proportionally related to the volume of ice; this ice volume effect is primarily a response to waxing and waning of continental ice sheets of the Northern Hemisphere and the Arctic Ocean Ice Sheet (Hughes et al. 1977; Shackleton 1977; Williams et al. 1981; among others).

2. Physical circulation processes are such that the oceanic reservoir is completely mixed in approximately 1000 years (Craig and Gordon 1965). Changes in the volume of continental ice therefore affect the isotopic composition of the oceans rapidly, nearly uniformly, and nearly synchronously within the mixing time of the oceans. These facts enable oxygen isotope records to be used in global and geologically synchronous correlations, especially when combined with geomagnetic and biostratigraphic data (see this volume, chaps. 1 and 2).

3. Regional differences in the evaporation-precipitation ratio of a basin (Deuser et al. 1976) or meltwater run-off into marginal seas like the Gulf of Mexico, Mediterranean Sea and the Labrador Sea will sometimes lead to isotopic changes much larger than those observed in the open ocean (Kennett and Shackleton 1975; Emiliani et al. 1975; 1977; Leventer et al. 1982; Williams et al. 1977; Cita et al. 1977; Vergnaud-Grazzini et al. 1977; Williams and Thunell 1979). As will be discussed later, isotopic events due to regional oceanographic differences very often enhance the stratigraphic and paleoenvironmental information contained in isotopic records from marginal seas.

4. A large portion of the oxygen isotope signal recorded in foraminifera (and other forms of biogenic carbonate) from marine sediments is due to changes in the isotopic composition of the oceans. Temperature changes coincident with glacial-interglacial climatic variations are only a small part of the Pleistocene isotope signal (typically less than 0.5 ‰ or 25% of the total glacial-interglacial signal).

5. The variables such as depth of growth, environmental tolerances, and various vital effects that may influence the isotopic composition of foraminifera are sufficiently well understood to

INTERGLACIAL CONDITIONS

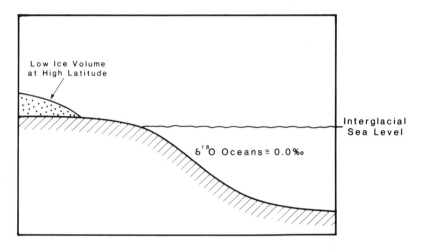

GLACIAL CONDITIONS

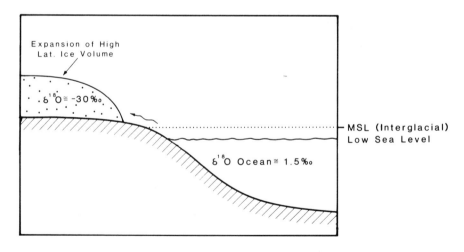

Figure 3.7 *A schematic representation of the approximate change in the oxygen isotopic composition of the oceans as a function of lowered sea level during glacial periods and the storage of the water as ice in high latitudes. The removed water is preferentially depleted in oxygen-18. The return of the isotopically negative water from continental high latitude ice sheets and glaciers results in (1) a return to interglacial sea levels, (2) adjustment of the isotopic composition of the ocean to interglacial $\delta^{18}O$ values as well as (3) negative meltwater anomalies in oxygen isotope records from many marginal seas.*

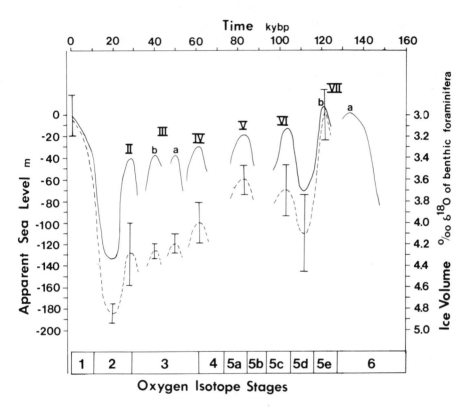

Figure 3.8 *A representation of the relationship between the composite oxygen isotopic record of deep sea benthic foraminifera for the last 130,000 (dashed line) and the glacio-eustatic sea level record (solid line) as determined by Moore (1982). The apparent differences between the oxygen isotopic and sea level records have been interpreted in terms of bottom water temperature change (Dodge et al. 1983), surface water temperature change (Aharon 1983) and the presence of substantial quantities of ice as floating marine ice sheets in the Arctic Ocean (Broecker 1975; Williams et al. 1981). The chronology for the sea level record is constrained by uranium disequilibrium dating of coral terraces. (From Williams et al. 1981. Reprinted by permission of Elsevier Scientific Publishing Company.)*

enable the confident use of foraminifera in oxygen isotope stratigraphy (Hecht 1976; Duplessy 1978; Vincent and Berger 1981; among others). The relationships defined by isotope studies of living and Recent fossil foraminifera have not changed in the time period represented by the Pleistocene (Savin and Douglas 1973; Williams and Healy-Williams 1980; Vincent et al. 1981; among others). Sufficient numbers of foraminiferal species (both benthic and planktonic) are available from most types of marine sediments for isotopic analyses.

6. The $^{18}O/^{16}O$ ratio in the calcite of fossil foraminifera is not affected by isotopic exchange with bottom or pore waters. Although diagenetic alteration due to recrystallization and severe dissolution may affect the oxygen isotope data (Baker et al. 1982; Elderfield et al. 1982; Killingley 1983), diagenesis is usually negligible in sediments of Pleistocene age, and potential effects are easily anticipated by microscopic examination of the foraminiferal tests.

7. Glacial to interglacial changes in the isotopic composition of the oceans have occurred quasiperiodically throughout the Pleistocene. These glacial-interglacial cycles can be identified as *isotope stages*. A total of 19 isotope stages occur from the present to 728,000 years ago at the magnetic polarity boundary between the Matuyama and Brunhes Epochs (Shackleton and Opdyke 1973; 1976; Emiliani 1978). Significant isotope events are apparent in earlier parts of the Pleistocene (van Donk 1976; Shackleton and Opdyke 1976; among others) and attempts to define isotope stages in the early Pleistocene will be discussed in the next section.

8. The oxygen isotope stages show unique characteristics in their isotope signals that enable them to be recognized, regardless of the ocean basin from which the sediments are recovered.

9. The chronology of the isotope stage boundaries has been firmly established by using (1) radiometric dating with ^{14}C, (2) sediment accumulation rate determinations from paleomagnetic stratigraphy (see chap. 1 this volume) and the decay of ^{230}Th and ^{231}Pa, and (3) spectral analysis of the isotope signal and correlation with the mathematically determined orbital parameters of the Earth (Hays et al. 1976; Kominz et al. 1979; Imbrie et al. in press; among others).

Description of ^{18}O Records for the Pleistocene

Figure 3.9 contains two composite oxygen isotope records for approximately the last 730,000 years. These are based on ^{18}O data from foraminifera from several Caribbean cores (Emiliani 1978) and from five cores distributed in different ocean basins (Imbrie et al., in press). Both isotope records show how the isotope signal represents quasi-periodic changes between ^{18}O-enriched intervals (positive glacial stages) and ^{18}O-depleted intervals (negative interglacial stages). These glacial-interglacial fluctuations represent changing climatic regimes with a dominant period of approximately 100,000 years. Emiliani (1955, 1966) established the convention by which interglacial isotope stages are odd-numbered and glacial isotope stages are even-numbered (fig. 3.9). Chronological differences exist between the two records, because of the methods used to assign age estimates to the stage boundaries. Emiliani (1978) primarily used linear interpolations between dated biostratigraphic horizons. The SPECMAP chronology (table 3.2) is based on using a combination of well-dated biostratigraphic datums and also tuning the record to the frequencies of the Earth's orbital parameters. As was mentioned in the preceding section of this chapter, spectral analysis of oxygen isotope records such as this from several ocean basins has revealed that the isotope signal contains significant power at the frequencies of 100,000; 41,000; and 23,000 years (Hays et al. 1976; Pisias and Moore 1980; Kominz et al. 1979). The chronology and character of the composite isotope record shown in the left column of figure 3.9 will likely become the standard for future stable isotope studies of this time period.

The glacial-interglacial isotope fluctuations seen in the late Pleistocene are now known to have occurred throughout the entire Pleistocene Epoch (about the last 1.8 million years). Largely as the result of the Deep-Sea Drilling Project (DSDP) oxygen isotope records for nearly the entire Pleistocene are available from the equatorial Atlantic (Van Donk 1976; Shackleton and Boersma 1979); the South Atlantic (Shackleton and Hall 1983; Vergnaud-Grazzini et al. in press); the equatorial Pacific (Shackleton and Opdyke 1976); and the western Caribbean (Prell 1982). By far the best isotope records for the entire Pleis-

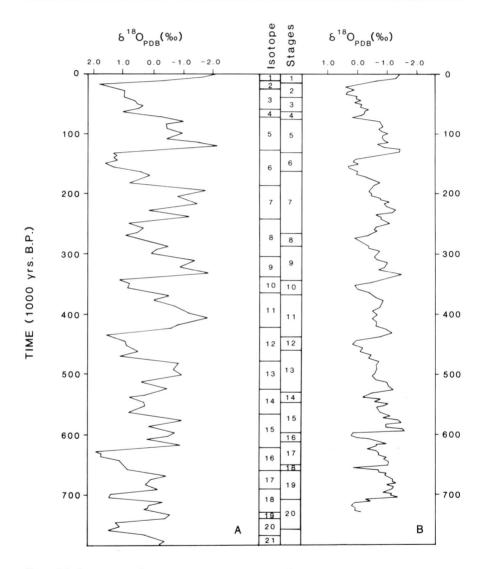

Figure 3.9 *A comparison of two composite oxygen isotopic records for the last 700,000 years (B taken from Emiliani 1978, right-hand column, and A from the SPECMAP group, Imbrie et al., in press, left-hand column). The differences in chronology for the oxygen isotope stages represent refinements of the chronology for the stage boundaries based on tuning the oxygen isotope record to the spectral frequency contained in the earth's orbital parameters (refer to the text for a more detailed discussion).*

tocene are from Pacific piston core, V28–239 (fig. 3.10) (Shackleton and Opdyke 1976) and from hydraulic piston cores from the Caribbean DSDP Site 502B (Prell 1982) (fig. 3.11).

As is shown in figures 3.10 and 3.11, quasi-periodic isotopic cycles exist throughout the Pleistocene, offering the potential stratigraphic resolution found in few other parameters. Glacial advances and glacio-eustatic sea-level changes on the order of those observed in the last 150,000 years appear to have characterized the isotope record of the last 800,000 years. The records also contain evidence for two different modes of climatic change. Glacial-interglacial $\delta^{18}O$ changes in the early Pleistocene (1.8–0.95 MYBP) became significantly different in frequency and amplitude from $\delta^{18}O$ changes in the late Pleistocene (Brunhes paleomagnetic Chron, 0.73 MYBP to present) (figs. 3.11 and 3.12). Early Pleistocene isotope records are characterized by lower-amplitude, but higher-frequency, glacial-interglacial fluctuations, with periods on the order of 41,000 years (Van Donk 1976; Shackleton and Opdyke 1976; Prell 1982; Pisias and Moore 1980). In the planktonic isotope record of Pacific core V28–239, for example, the mean $\delta^{18}O$-values in the late Pleistocene (fig. 3.12) are significantly heavier (enriched in ^{18}O), and the average isotopic change between individual glacial-interglacial episodes ($\Delta^{18}O$) is more variable (0.90 0.27 ‰ for 1) than in the early portion of the Pleistocene, in which the mean glacial-interglacial change ($\Delta^{18}O$) is only 0.15 (1σ) with an overall mean of 0.60 ‰. Even though sediment accumulation rates in core V28–239 are extremely low, the character of its Pleistocene isotope record agrees well with new isotopic evidence from a continuous hydraulic piston–core (Site 502B) from the Caribbean, with a higher accumulation rate (Prell, 1982). The Pleistocene record of DSDP Site 502 (fig. 3.11) shows evidence for the same two climatic regimes as piston core V28–239. The early Pleistocene mode of variation shows high-frequency isotopic oscillations of 0.28 ‰ (1σ) about a mean of −0.82 ‰. Just after the Jaramillo Subchron (0.9 MYBP) (see chap. 1), the glacial-interglacial isotopic oscillations change to higher amplitude (but lower frequency) variations of 0.50 ‰ (1σ) about a mean of −0.46 ‰ in the Pleistocene.

This systematic shift in average $\delta^{18}O$ values between the Lower and Upper Pleistocene can be seen in a comparison of oxygen

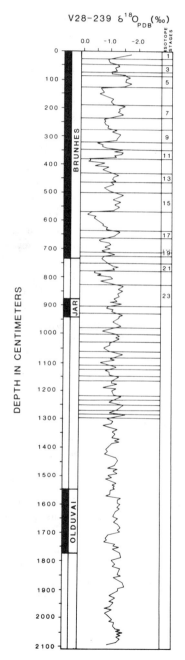

Figure 3.10 *The oxygen isotope record for the entire Quaternary from Pacific core V28–239 by Shackleton and Opdyke (1976). The oxygen isotope stages are defined up to stage 23 at the Jaramillo subchron in the Matayuma chron. Shackleton and Opdyke (1976) attempted to show the potential for isotope stages below the Jaramillo subchron but did not propose formal stage designations at that time. The paleomagnetic stratigraphy and oxygen isotope record are adapted from Shackleton and Opdyke (1976).*

Figure 3.11 *The detailed oxygen isotope record for Caribbean DSDP Site 502B representing the first record for nearly the entire Pleistocene (Prell 1982) from a marine sequence obtained with the newly developed hydraulic piston core (Chapter 4). The oxygen isotope record, paleomagnetic stratigraphy, and the Ericson* Globorotalia menardii *biostratigraphic zones for Site 502B are taken from Prell (1982).*

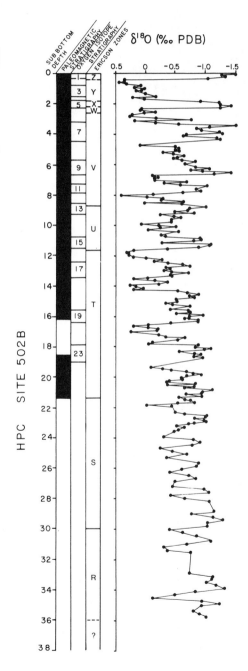

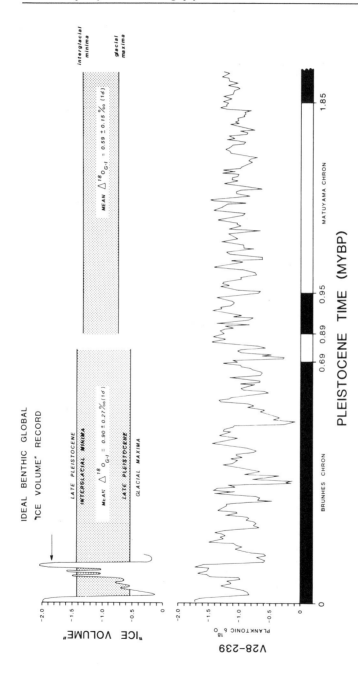

Figure 3.12 *Pleistocene oxygen isotope record for equatorial Pacific core V28–239 showing the statistical difference in both the mean interglacial minima and glacial maxima, and the mean glacial-interglacial isotope change ($\Delta^{18}O_{G-I}$) between the late Pleistocene (Brunhes Chron to present) and early Pleistocene (Upper Matuyama Chron, 1.8 to 0.95 MYBP). High-frequency, low-amplitude isotopic changes characterize the early Pleistocene, and low-frequency, higher-amplitude isotopic changes occur in the late Pleistocene. An idealized benthic isotope record from high accumulation rate cores is shown for comparison in the upper left-hand corner of the figure. (From Williams et al. 1981. Reprinted by permission of Elsevier Scientific Publishing Company.)*

isotope records from various oceans, including the Mediterranean Sea (fig. 3.13) (Thunell and Williams 1983). The mid-Pleistocene isotope shift appears to be global in nature and is recorded by the $\delta^{18}O$ of both benthic and planktonic foraminifera. Beside individual isotope stages, therefore, the observed long-term trends in the Pleistocene oxygen isotope record also can be of use in stratigraphic correlation and paleoclimatic inferences of Pleistocene marine sections. As is shown in figure 3.13, strong evidence exists for earlier systematic climatic shifts in the Pliocene, and these shifts may be of stratigraphic as well as paleoclimatic importance (Thunell and Williams 1983). Besides determining the exact significance of these climatic shifts in the isotope record, however, significant work also remains to be done in accurately establishing a chronology for the early Pleistocene isotope variations.

No attempt has been made in figures 3.9 and 3.10 to extend the numbering of the isotope stages beyond isotope stage 23 (about 1 MYBP) into the early Pleistocene. Van Donk (1976) had earlier attempted extending the isotope stages into the early Pleistocene by using the isotope record of V16–205, but he was hampered by the low accumulation rate (0.5 cm/10^3 yr.) in this Atlantic core (fig. 3.14). These early Pleistocene "stages," as designated here, should be regarded as entirely preliminary, until more isotope records of suitable quality become available. The author feels, however, that such an attempt still can prove useful, as will be shown in the next section.

Oxygen Isotope Records from the Gulf of Mexico

This section contains examples of some published and unpublished oxygen isotope records from Gulf of Mexico sediments. In general, very few oxygen isotope records are available from the Gulf of Mexico as compared to isotope records from the major ocean basins and other large marginal seas like the Mediterranean. As will be shown, however, enough records are available from the Gulf of Mexico to enable us to use isotope records confidently for stratigraphic correlation of cores and offshore

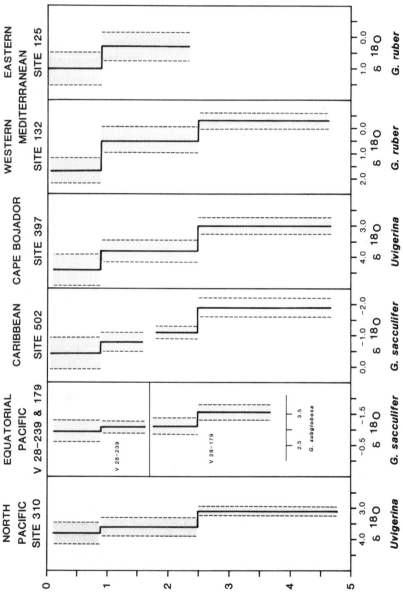

Figure 3.13 Schematic diagram comparing the mean oxygen isotopic values and their variability in δ¹⁸O records for DSDP Site 310 from the North Pacific (Keigwin 1979), equatorial Pacific cores V28–239 and V28–179 (Shackleton and Opdyke 1976 and 1977), DSDP Site 502 from the Caribbean (Keigwin 1982; Prell 1982), DSDP Site 397 from Cape Bojador (Shackleton and Cita 1979), and DSDP Sites 125 and 132 from the Mediterranean (Thunell and Williams 1983). The vertical lines denote the mean δ¹⁸O value for the time period representing significant shifts or steps in climatic conditions. The stippled area represents ± 1 standard deviation about the mean. (From Williams and Thunell 1983. Reprinted by permission of the editor of the Utrecht Micropaleontological Bulletins.)

Figure 3.14 *The Pleistocene oxygen isotope record of Atlantic core V16–205 adapted from Van Donk (1976), showing his attempt to extend the designation of the oxygen isotope stages beyond isotope stage 23 to include stages up to isotope stage 39 in the Olduvai Subchron of the reversed Matuyama Chron. The paleomagnetic stratigraphy and biostratigraphy are from Briskin and Berggren (1975).*

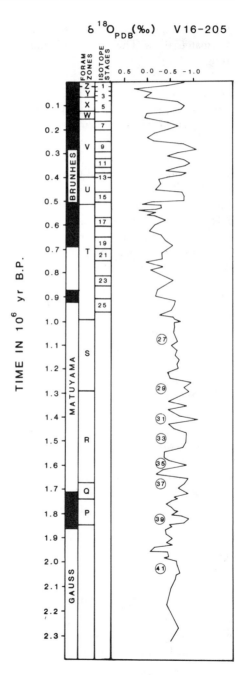

wells, as well as to make paleoenvironmental interpretations of this marginal sea. The location of many of these records is shown in figure 3.15.

Record of the Last 30–40 Thousand Years

The first oxygen isotope work in Gulf of Mexico deep-sea sediments yielded dramatic results (Kennett and Shackleton 1975; Emiliani et al. 1975). The latest Pleistocene portion of the isotope record for the Gulf not only portrayed the global isotopic change from the last glacial maximum to the Holocene interglacial, but also contained several unique characteristics. For example, the $\delta^{18}O$ record from the western Gulf of Mexico (fig. 3.16) shows the expected 1.5 ‰ global change from isotope stage 2 (the last glacial maximum) to isotope stage 1 (the present Holocene interglacial). Note, however, the very negative values recorded at approximately 13 KYBP (*shaded area* in fig. 3.16). This unique isotope record results from the fact that the Gulf of Mexico (1) is a semirestricted marginal sea contiguous to continental North America and (2) directly receives a large portion of freshwater drainage via the Mississippi drainage system and thus has had an intimate relationship with the history of the Laurentide ice sheet. The water locked as ice in the Laurentide ice sheet was very depleted (isotopically negative) in ^{18}O. Owing to its connection with the Mississippi drainage system, the Gulf is the first basin to receive the melt products from the midcontinental portions of the ice sheet as it melts and returns during the climatic amelioration at the end of glacial episodes. Therefore, in addition to the global ice-volume signal, the isotope record of the Gulf was strongly influenced by the influx of isotopically light meltwater (^{16}O-enriched) at the end of the last major continental glaciation (Kennett and Shackleton 1975; Emiliani et al. 1975; 1977; Leventer 1981). The influx of glacial meltwater from the Laurentide ice sheet is recorded as unusually negative $\delta^{18}O$ values superimposed on the global ice-volume signal (fig. 3.16). As will be discussed later, the surface waters of the Gulf have been periodically impacted by this meltwater at the end of each major continental glaciation, which characterized the Pleistocene, and that these meltwater events offer additional information for making stratigraphic correlations as well as for understanding the depositional history of the thick Pleistocene sections along the northern continental margin of the Gulf of Mexico.

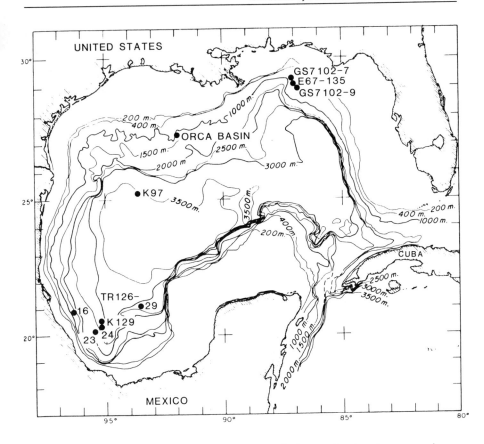

Figure 3.15 Bathymetric map of the Gulf of Mexico showing the location of various piston cores and boreholes ("Eureka" 67–135) for which oxygen isotope records will be illustrated and discussed in the remainder of the text.

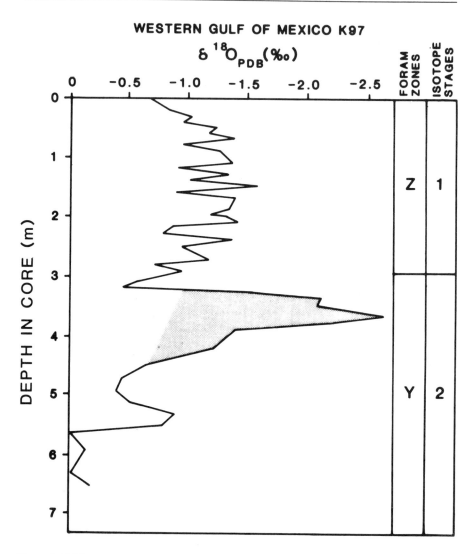

Figure 3.16 *The oxygen isotope record for piston core K97 from the western Gulf of Mexico (fig. 3.15) for approximately the last 25,000 years showing the negative oxygen isotope event (shaded area of curve) which Kennett and Shackleton (1975) inferred to reflect the input of isotopically depleted glacial meltwater from the Laurentide ice sheet during the deglaciation between isotope stage 2 and 1 (adapted from Kennett and Shackleton 1975).*

The recovery of anoxic sediments from the intraslope Orca Basin located on the continental slope of the Gulf of Mexico has now made it possible to examine the nature of the meltwater influx without the blurring effects of bioturbation by bottom-dwelling organisms. The Orca Basin is a 400 km^2 depression on the continental slope, 225 km south of the Louisiana coast (fig. 3.15) and is the only presently known anoxic basin among the many intraslope basins of the northwestern Gulf of Mexico (Shokes et al. 1977; Tompkins and Shepard 1979; Trabant and Presley 1978; Bouma et al. 1978). The basin is ideally located to have recorded meltwater discharge from the Laurentide ice sheet. The anoxic conditions beneath the hypersaline brine of the basin have led to the deposition of laminated, organically rich, un-bioturbated sediments, because benthic life is almost completely absent. Oxygen isotopic analyses performed on a surface-dwelling species of planktonic foraminifera (*Globigerinoides ruber*, white variety) from one such piston core (EN32–PC6) reveal a detailed history of the Laurentide deglaciation process over the last 29,000 years (fig. 3.17) (Leventer 1981; Leventer et al. 1982). A detailed record is permitted in the sediments of the Orca Basin, owing to the relatively large sediment accumulation rates (29 to 49 cm/1000 yr.), which provide a sampling interval equivalent to about every 250 years throughout the entire core and about every 125 years throughout the period of greatest meltwater influx.

The oxygen isotope record of EN32–PC6 (fig. 3.17) shows that the glacial maximum conditions occurred between 17.75 and 16.5 KYBP, with the initial introduction of isotopically depleted glacial meltwater beginning at 16.5 KYBP. This first major discharge of meltwater was comprised of two short pulses lasting from 16.5 to 15.5 KYBP, and 15.25 to 14.9 KYBP, showing magnitudes of 2 ‰ and 0.5 ‰, respectively. The second major discharge lasted from 15 to 11.6 KYBP, with a total magnitude of 2.6 ‰. Six shorter meltwater pulses comprise this phase of the discharge, with peak depletions of from 0.7 to 1.4 ‰. These spikes may also, in part, represent temperature variability superimposed on the meltwater signal. The total meltwater anomaly, as measured from 16.5 to 13.1 KYBP, shows a magnitude of 3.6 ‰ and is superimposed upon a change of about − 1.7 ‰ from average glacial isotope values at the end of the last glaciation (29 to 18 KYBP) to interglacial values of the Holocene (9–0 KYBP).

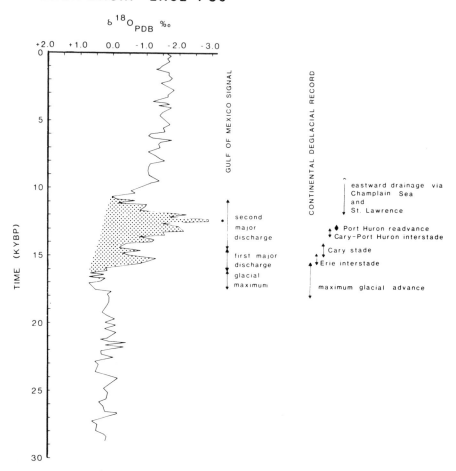

Figure 3.17 *The detailed oxygen isotope record from piston EN32–PC6 recovered from the anoxic Orca Basin (fig. 3.15) illustrating the most detailed record available to date of the oxygen isotopic anamolies associated with episodic influx of meltwater into the Gulf of Mexico at the last deglaciation. A potential correlation between the meltwater record and events of the continental deglacial record is also shown (adapted from Leventer 1981, and Leventer et al. 1982).*

A comparison of the latest Pleistocene oxygen isotope record from the Orca Basin with records from other Gulf of Mexico cores (Emiliani et al. 1977, Kennett and Shackleton 1975; and Falls 1980) reveals an overall similarity, but with some important differences (fig. 3.18). In all four cores, meltwater discharged to the Gulf of Mexico commenced between 18 and 16.5 KYBP and ended between 12 and 10.5 KYBP. Maximum isotopic depletion occurred between 12 and 14 KYBP in all four cores, indicating that meltwater influx to the Gulf was most intense at this time.

Details in the Orca Basin isotope curve are not evident in the other three Gulf of Mexico cores, however, because of the lower sediment accumulation rates combined with wider sampling intervals in the other cores. Differences in the magnitude of the meltwater anomaly among the isotope records are most probably the result of varying distances from the mouth of the Mississippi River, the conduit through which meltwater emptied into the Gulf (fig. 3.15). A plot of the $\delta^{18}O$ value at the zenith of the meltwater influx in each core versus the approximate distance that each core is located from the Mississippi discharge (fig. 3.19) shows (1) lighter, more negative $\delta^{18}O$ values closer to the source of the meltwater and (2) a mixing line of more positive $\delta^{18}O$ values with increased distance from the point of meltwater discharge, as the isotopically depleted meltwater mixes with the saline surface waters of the Gulf. This observation is particularly important because it describes how the meltwater isotopic signal is attenuated with distance from the source and will aid in the interpretation of older isotope records from the Gulf in terms of earlier glacial, interglacial and meltwater episodes.

Late and Early Pleistocene Isotope Records from the Gulf of Mexico

In addition to this interest in the isotope record of the last 30,000 years, later work by Malmgren and Kennett (1978) showed that a positive correlation existed between the oxygen isotope record for approximately the last 200,000 years and a relative paleotemperature curve for a core (K129) from the western Gulf of Mexico (fig. 3.20). The first principal component of the total planktonic foraminiferal fauna was used as the relative paleotemperature curve (see chap. 2, this volume). The major isotope stages 1–6 are clearly discernible, and correlatable with the Ericson bio-

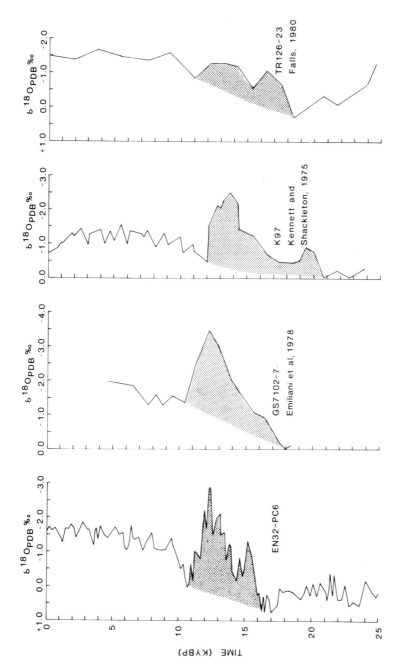

Figure 3.18 A comparison between four Gulf of Mexico oxygen isotope records spanning the last 25,000 years, illustrating the magnitude and approximate timing of the meltwater event in the Gulf of Mexico (stippled region in each record). Refer to figure 3.15 for the location of the cores. (From Leventer et al. 1982. Reprinted by permission of Elsevier Scientific Publishing Company.)

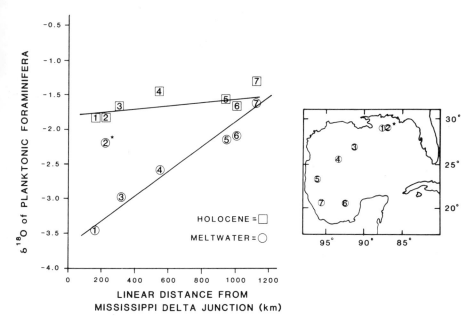

Figure 3.19 *A plot representing the change in isotopic gradient of planktonic foraminifera with distance from the Mississippi Delta junction during the Holocene (square symbols) and at the time of the maximum extent of the meltwater anamoly associated with the last deglaciation (circled symbols). The difference in slope between the Holocene and meltwater δ^{18}O gradients shows that the magnitude of the isotopic anamoly is largest near the source of the meltwater and decreases linearly away from the source (a strong positive correlation with an r[2] value = +0.92 if the one exception to the meltwater gradient, seen in core 2 (GS7102–09) is eliminated from the least squares regression). Core 2 is located farthest to the east of the other cores, and we believe the lack of a strong meltwater anomaly in this core results the westward counterclockwise flow of the meltwater along the Louisiana-Texas margin. Data from each of the cores were taken from the following: core 1, GS7102–07 and core 2, GS7102–9 from Emiliani et al., 1975, 1977; core 3, EN32–PC6 from the Orca Basin from Leventer et al. 1982; cores 4 (K97), 5 (K120), and 6 (K139) from Kennett and Shackleton 1975, and core 7, TR126–23, from Falls 1980.*

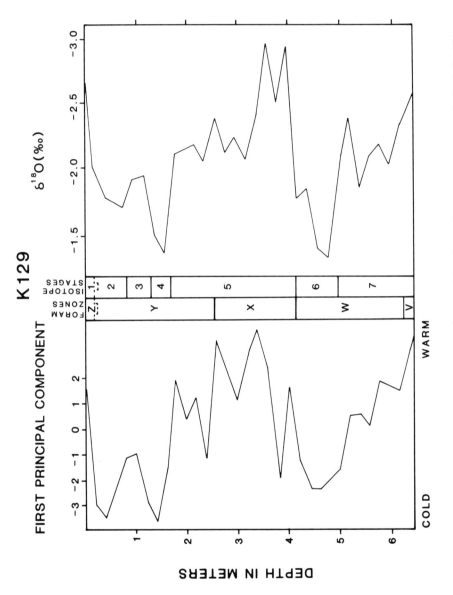

Figure 3.20 *A comparison between the Ericson biostratigraphic zones (chap. 2) and the oxygen isotope stratigraphy for the late-Pleistocene Gulf of Mexico core K129 (adapted from Malmgren and Kennett 1978) (refer to fig. 3.15 for location of K129).*

stratigraphic zonation. The foraminiferal zone Z is approximately equivalent to isotope stage 1, foraminifera zone Y is subdivided by isotope stages 2, 3, and 4 and foraminifera zone X is equivalent to a large portion of isotope stage 5. However, it is not possible clearly to determine the substages of stage 5 (a–e), owing to the relatively large sample spacing used in the study of K129. The foram zone W, like the Y zone, extends across the isotope stage 6/7 boundary and includes better than the upper half of isotope stage 7 (fig. 3.20).

A more detailed $\delta^{18}O$ record for the late Pleistocene is shown for core TR126–23 from the southwestern Gulf (figs. 3.21, 3.15) (Williams and Fillon, unpublished data). On the basis of detailed analyses of *G. ruber* at sample intervals equivalent to about 1000 years, the isotope stages 1–6 and substages of stage 5 are clearly defined in the record. The core therefore contains no hiatuses and has an average sediment accumulation rate of nearly 5 cm/ 1000 yr. Detailed records such as this contain valuable information regarding the timing of meltwater events and regional paleoenvironmental changes. The $\delta^{18}O$ record also permits the placement of the Ericson biostratigraphic zones (see chap. 2) and volcanic tephra layers, Y8 and W1 (see chaps. 1 and 4), within an absolute time framework. The Y/Z boundary occurs just above the meltwater events (identified by the bold arrows in fig. 3.21) very near the isotope stage 2/1 boundary. The X/Y boundary, along with the *Globorotalia flexuosa* extinction datum, falls within isotope substage 5b, and the W/X boundary is placed near the isotope stage 6/5e transition (fig. 3.21). The largest glacial-interglacial $\delta^{18}O$ signal (2.7 ‰) is recorded from stage 6 to stage 5, and is clearly greater than the maximum range (1.8 to 2.0 ‰) observed in most open-ocean cores (figs. 3.3, 3.6). This enhanced signal is due to the earlier discharge of isotopically light meltwater at the end of the glacial episode represented by isotope stage 6 (*bold arrow* in fig. 3.21).

Figure 3.22 shows an effort to document this earlier meltwater signal in other cores that contained sediment spanning the isotope stage 6/5 boundary (from 100 to 150 KYBP) (Falls, 1980). Although the major portion of the meltwater signal is associated with the isotope stage 2/1 boundary (figs. 3.17, 3.18), the meltwater signal at the end of isotope stage 6 appears to be "welded" onto substage 5e, producing a large isotopic range but no clearly

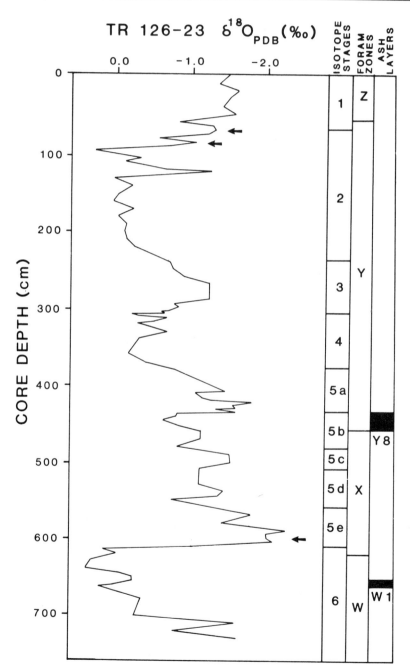

Figure 3.21 *The detailed oxygen isotope record for the last 140 KYBP for Gulf of Mexico core TR126–23 (fig. 3.15) in comparison with the Ericson biostratigraphic zones and the presence of the Y8 and W1 tephra layers (Ledbetter, Chapter 4). The bold arrows near the oxygen isotope stage 2/1 boundary and near the isotope stage 6/5e boundary illustrate the input of isotopically depleted meltwater at glacial termination I and II, respectively (Williams and Fillon, unpublished data).*

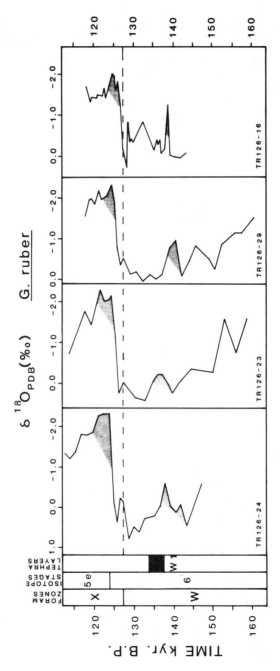

Figure 3.22 Oxygen isotope records from four southwestern Gulf of Mexico cores (TR126–16, –23, –24, –29; fig. 3.15) across the isotope stage 6/5e boundary showing (1) the meltwater anomaly associated with stage 5e (shaded area) and (2) the occurrence of the W1 tephra layer in the negative isotopic event in the late part of isotope stage 6 (adapted from Falls 1980).

definable meltwater "event." No explanation is offered at this time for this difference. Falls (1980) also consistently found the W/X biostratigraphic boundary at the base of the stage 6/5e transition and the W1 tephra layer in isotope stage 6 associated very closely with a slight negative isotope signal (fig. 3.22).

The only published oxygen isotope record for Gulf of Mexico sediments older than the last 150,000 years comes from the biostratigraphic and isotopic study of Eureka 67–135 (fig. 3.23) drilled for Shell Oil Co. in a water depth of 725 m in the DeSoto Canyon (Brunner and Keigwin 1981). As is shown in the right-hand column of figure 3.23, the preliminary foraminiferal biostratigraphy indicated a relatively complete stratigraphic section in the Pleistocene section of E67–135 (Brunner and Keigwin, 1981). Comparison of the E67–135 isotope record with any other Pleistocene isotope record (for example, refer to fig. 3.10 or 3.11) suggests that numerous isotope stages may be missing in E67–135. For example, stage 5 is closely associated with zone X (fig. 3.21) and at least four quasi-periodic isotopic oscillations (from stages 13 to 12, 11 to 10, 9 to 8 and 7 to 6) with an amplitude of ~1 ‰ should be recorded if complete recovery had occurred in the biostratigraphic zone V (compare the V interval of Caribbean Site 502B with that of E67–135) (figs. 3.11 and 3.23; table 3.3).

The same lack of definition of isotope stages occurs in earlier portions of the E67–135 record, although the isotopic shift toward heavier average $\delta^{18}O$ values from the early to late Pleistocene (below and above the middle of zone T may be discernible. The lack of the appropriate number of isotopic oscillations with characteristic glacial-interglacial amplitudes might suggest the existence of numerous breaks in sedimentation in the Pleistocene section of E67–135.

To further address this question, detailed stable isotopic (Williams, unpublished data) and biostratigraphic (Neff 1983; chap. 2, this volume), studies were undertaken with the samples utilized by Brunner and Keigwin (1981) and augmented with samples at closer intervals provided by Stephan Gartner of Texas A&M University. The new stable isotope results (fig. 3.24) are based on analyses of the surface-dwelling species, *G. ruber*. Although the analyses have not been completed on all available samples at present, even a preliminary comparison of the benthic and planktonic isotope records indicates that the planktonic $\delta^{18}O$

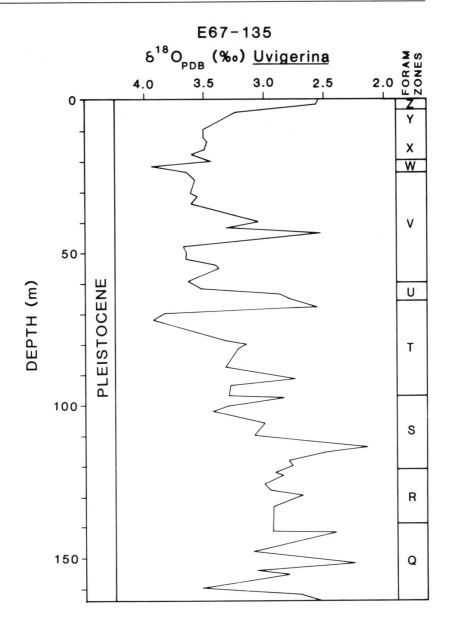

Figure 3.23 *The Pleistocene oxygen isotope record and foraminiferal biostratigraphy from the drilled section of Eureka 67–135 (fig. 3.15). (Adapted from Brunner and Keigwin 1981.) The* $\delta^{18}O$ *record is based on analyses of specimens of the benthic foraminiferal genus,* Uvigerina.

Ericson Zonation	Isotope Stages	No. of Glacial-* Interglacial Cycles	Equivalent N. American Stages***
Z	1/2	–	Holocene
Y	2,3,4	2	
X	4/5–5/6	2**	Wisconsinan
W	6	–	
V	6–12/13	3	Sangamon
U	13–15/16	2	Illinoisian
T	16–25	3	Yarmouth
S	(25–38)	6–7	Kansan
R	(38–43)	2–3	Nebraskan

Table 3.3 *Comparison Between the Ericson Biostratigraphic Zones, Oxygen Isotope Stages, and Classical North American Glacial Stages of the Pleistocene*

* $\delta^{18}O$ change between glacial to interglacial $\cong$ 1 to 0.5 ‰
** Substages of 5d–5c, 5b–5a
*** Fillon, chap. 5 (this volume)
Note: Isotope stage numbers in parentheses should be regarded as preliminary.

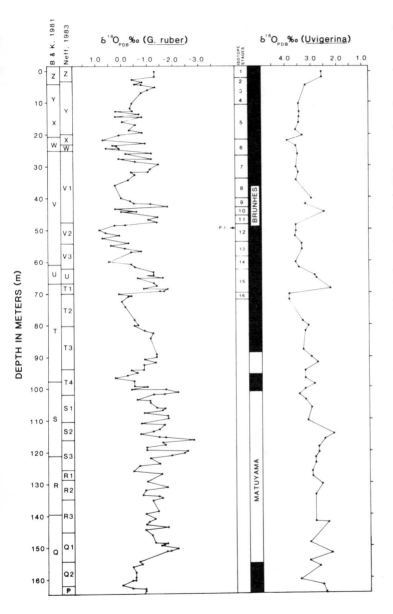

Figure 3.24 *Unpublished oxygen isotope data from E67–135 based on analyses of the planktonic foraminifera, Globogerinoides ruber, and compared with the δ¹⁸O data on the benthic foraminifera Uvigerina from Brunner and Keigwin (1981). An attempt has been made to identify the equivalent oxygen isotope stages in the planktonic record. The stratigraphic occurrence of the last appearance datum of the coccolith Pseudoemiliania lacunosa, as identified in E67–135 by Chen (1978), occurs in the upper part of isotope stage 12 in agreement with the work of Thierstein et al. (1978) (fig. 3.27). Also shown for comparison with the planktonic δ¹⁸O record are (1) the Ericson biostratigraphic zonations of Brunner and Keigwin, (1981) and Neff (1983), and (2) the inferred paleomagnetic polarity boundaries.*

record appears more completely to describe the oxygen isotope stages expected from the isotope stratigraphy for the Pleistocene (Shackleton and Opdyke 1976; Van Donk 1976; Prell 1982).

The glacial-interglacial changes in the late Pleistocene portion of the planktonic isotopic record are as large as 3.0 ‰, as seen for example, the upper portion of foraminifera zone V from 35 to 30 meters subbottom depth and 1.9 ‰ in the upper portion of the foraminifera zone T from 70 to 65 m subbottom depth (fig. 3.24). The latter large isotopic event in the planktonic record (possibly isotope stage 7 in foraminifera zone V) is not seen in the benthic record, owing to sample spacing; however, the very light values which are probably all or part of isotope stages 15 and 17 are reflected in both the benthic and planktonic isotope records (fig. 3.24) across the interval identified as zones U and T1 by Neff (1983). The compressed extent of the U zone in E67–135 (fig. 3.24) is likely the result of a hiatus in sedimentation, because, as was seen at Caribbean Site 502B (fig. 3.25), zone U should contain the stage 15/16 boundary, stages 15, 14, and most of 13. The $\delta^{18}O$ values in the zone U of E67–135 do not define clearly these stages and are more negative than -1.5 ‰, which is a characteristic value for interglacial conditions in this part of the Gulf of Mexico (see figs. 3.18 and 3.21, for example).

As another example of the stratigraphic usefulness of isotope stratigraphy, the isotope record of E67–135 for the early Pleistocene zone S is examined and compared to the isotope record of Caribbean DSDP Site 502 (fig. 3.26). Although the (Illinoian) zone U contains three isotope stages in the Middle Pleistocene, zone S (Kansan) contains 12 to 13 isotope events in open-ocean records that may show stratigraphic significance as isotope "stages" (figs. 3.10, 3.11; table 3.3). An attempt to stratigraphically divide zone S by using the planktonic isotope record of E67–135 results in 11 to 12 significant isotope events, very similar to the number of "stages" found in the early-Pleistocene zone S of DSDP Site 502B (fig. 3.26). The isotope record of E67–135 thus indicates that the zone S of this borehole is relatively continuous, despite the uncertainties involved in actually assigned formal stage numbers to the isotope events of the early Pleistocene. When analyses of the additional available samples from E67–135 are completed, it will be possible to evaluate the stratigraphic completeness of the other sections of E67–135 in a similar man-

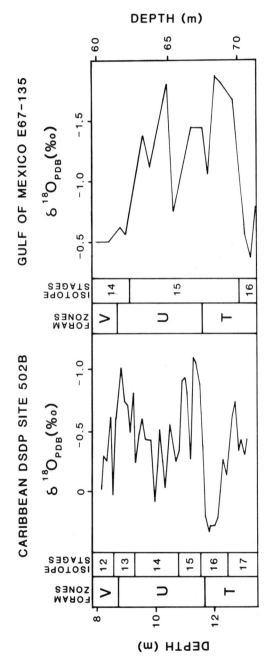

Figure 3.25 *A comparison of the oxygen isotope record of Caribbean core site 502B (Prell 1982; fig. 3.11) with that of E67–135 within the Ericson U biostratigraphic zone for each section.*

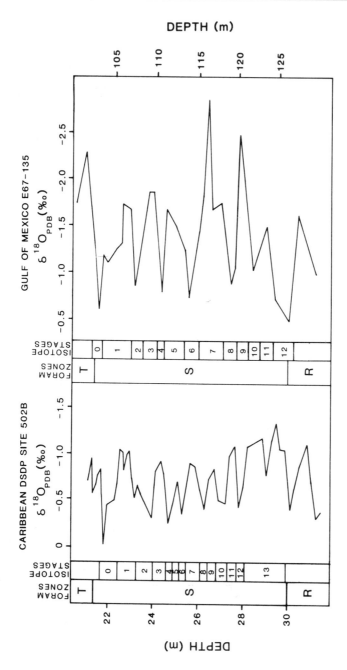

Figure 3.26 *A comparison of the oxygen isotope records for Caribbean site 502B and Gulf of Mexico E67–135 during the early Pleistocene (Kansan) S biostratigraphic zone. The ability to identify nearly the same number of isotope events in both records indicates that both these sections are represented by nearly equivalent stratigraphic recovery. The biostratigraphic zonation of Neff (1983) is shown in the E67–135 record. This figure illustrates how the oxygen isotope method can be used to study the stratigraphic continuity of early Pleistocene sections in the Gulf of Mexico. Additional sections spanning the early Pleistocene intervals could thus be correlated by comparing the number of isotope events found within these sections.*

ner. In fact, given the large sediment accumulation rates of the Gulf of Mexico, isotope studies of additional Gulf boreholes could contribute significantly toward refining the isotope states of the Middle to Early Pleistocene.

The δ^{18}O record of planktonic foraminifera from Pleistocene Gulf of Mexico sections like E67–135 also may provide important information regarding the influx of meltwater evolving from continental ice sheets during continental glaciations earlier than the Wisconsinan. Using sediments from very near the location of E67–135 in the DeSoto Canyon by Emiliani et al. (1975, 1977), isotopic analyses of foraminifera indicate that δ^{18}O values for *G. ruber* exceeded values of 2.0 ‰ during the influx of isotopically depleted meltwater at the last deglaciation (figs. 3.18, 3.19). Examination of the *G. ruber* δ^{18}O data from E67–135 (fig. 3.24) shows that other times occurred during the Pleistocene when δ^{18}O values lighter than this 2.0 ‰ value were exceeded, indicating the influx of meltwater during earlier deglaciations. Other interglacial intervals of the Pleistocene record, however, indicate either that meltwater events were not recorded in the existing data set or that the magnitude of the meltwater influx varied with time. Completion of the remaining E67–135 samples should enable use of the oxygen isotope record to determine the meltwater history of the Gulf of Mexico during the Pleistocene and, in conjunction with the detailed biostratigraphy (Neff 1983; see also chap. 2), magnetostratigraphy (chap. 1 and chap. 4, tephrochronology) to evaluate the degree of stratigraphic completeness in the Pleistocene section of E67–135.

Prospects for Application of Oxygen Isotope Stratigraphy to Drilled Sections from Continental Margin Sequences

The character of the Pleistocene oxygen isotope record from the Gulf of Mexico is sufficiently well established to enable the application of oxygen isotope stratigraphy to continental margin sequences from the Gulf coast and other sedimentary basins. In turn, isotope studies of Gulf sections with high rates of sediment accumulation will help to refine the nature and timing of the isotope stages in the early Pleistocene. Our experience indicates that small (less than 15 cc) samples of bulk sediment from continental margin boreholes typically contain sufficient foraminifera for $\delta^{18}O$ analyses. It should therefore be possible to utilize samples taken from test wells and exploration holes by either sidewall coring or from cuttings, to make $\delta^{18}O$ stratigraphy a viable technique for correlation between sections. It can be estimated, for example, that in a well with a 10,000-ft. Pleistocene section, sampled at 60-ft. intervals, approximately 300 $\delta^{18}O$ analyses would provide the resolution necessary to define the presence or absence of oxygen isotope stages. To obtain sufficient analysis in some zones of suspected low foraminiferal abundance or to ensure against poor side-wall core recovery, it may be necessary to double-shoot some cores or to use cuttings larger than 15 cc. A high-resolution $\delta^{18}O$ stratigraphic study in overlapping Pleistocene sections would permit the establishment of a standard oxygen isotope "type section" for the Gulf of Mexico, particularly when combined with detailed biostratigraphic, tephrochronologic and paleomagnetic studies of the same section (see chap. 2 and chaps. 1 and 4). Paleo–sea-level estimates based on changes in ice volume could be inferred from positive $\delta^{18}O$ values in the $\delta^{18}O$ record. Depositional models for parts of the continental margin could be constructed within the glacio-eustatic sea level history of the region. Changes in depocenters along the continental margin could be detected and related to high or low sea-level stands. The linkage between augmented Mississippi discharge during periods of deglaciation and the lithology of continental margin–drilled sections could be determined. The oxygen isotope stratigraphy also would permit more accurate biostratigraphic and lithostratigraphic correlations. As is shown in figure 3.27, oxygen isotope records have been used to determine the timing of biostratigraphic datum levels in relation to the isotope stages and past climatic conditions. For example, two coccolith datum levels which are important in Pleistocene sediments (the first appearance of *Emiliania huxleyi* and the extinction of *Pseudoemiliania lacunosa*) can be shown to be globally synchronous (fig. 3.27). *P. lacunosa* becomes extinct in the middle of oxygen isotope stage 12, approximately 458,000 years BP, and *E. huxleyi* reached its first consistent appearance during isotope stage 8 approximately 268 KYBP (Thierstein et al. 1977). The extinc-

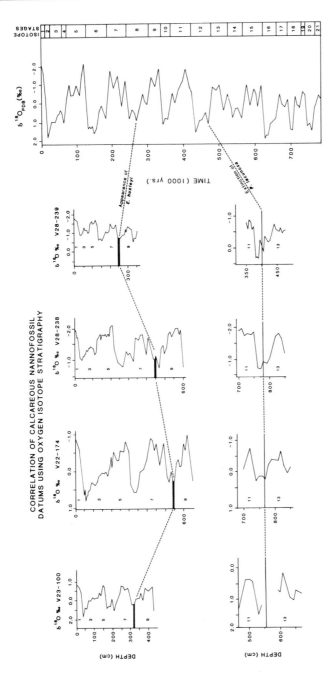

Figure 3.27 *The use of oxygen isotope stratigraphy to test the synchronous nature of biostratigraphic datums in calcareous nannofossils. The first appearance level of Emiliania huxleyi consistently appears in the middle part of isotope stage 8 (approx. 268 KYBP) in four cores from the Atlantic (V23–100; V22–174) and Pacific (V28–238; V28–239). The placement of the last appearance of Pseudoemiliania lacunosa occurs in the middle of isotope stage 12 approx. 458 KYBP (figure adapted from Thierstein et al. 1977). This approach has been used to determine the isochronous or diachronous nature of other local and regionally important micropaleontological tops in sedimentary basins like the Gulf of Mexico. Placement of the individual isotope records is shown with respect to the "stacked" composite isotope record from the SPECMAP group (Imbrie et al., in press).*

tion of *P. lacunosa* at 49 m subbottom depth in E67–135 (Chen 1978) occurs in isotope stage 12 as indicated by the isotope record of *G. ruber* (fig. 3.24). In other Gulf sections, $\delta^{18}O$ stratigraphy could be used to refine the stratigraphic usefulness of important biostratigraphic horizons in benthic and planktonic foraminiferal zonations and confirm the synchronous nature of microfossil datums within the framework of the Pleistocene isotope stages.

Summary

As was stated at the onset of this chapter, my objectives were to show how oxygen isotope stratigraphy can be applied to Gulf of Mexico sediments to provide quantitative, globally synchronous correlations within an absolute time framework. The stratigraphic completeness of a particular section of the Pleistocene can be tested by knowing the number of $\delta^{18}O$ cycles that should be present. The methodology has been perfected to the level whereby even early Pleistocene isotope stages with isotope changes greater than 0.5 ‰ can be resolved. Changes in sediment-accumulation rates or sediment type can be related directly to the Pleistocene glacial-interglacial cycles and glacio-eustatic sea-level changes. The character and chronology of the oxygen isotope stages are now established well enough through comparison with continuous deep-sea sedimentary cores, that it should now be possible to construct a standard, "Pleistocene $\delta^{18}O$-type section" for sedimentary basins like the Gulf of Mexico. By using such a standard $\delta^{18}O$-record, oxygen isotope stratigraphy could then be used to (1) enhance the stratigraphic resolution of continental margin drilled sections, (2) detect significant but short-duration hiatuses that are unresolvable using biostratigraphy and/or magnetostratigraphy alone, (3) determine unequivocally that certain biostratigraphic markers are indeed synchronous "datums" of chronostratigraphic significance, and (4) place lithostratigraphic correlations within the framework of global sea-level history (ice volume) and, as in the case of the Gulf of Mexico, meltwater discharges from the Mississippi River.

4

Late Pleistocene Tephrochronology in the Gulf of Mexico Region

Michael Ledbetter
Moss Landing Marine Laboratory
San Jose State University
Moss Landing, California   95039

4

Introduction

Volcanic ash layers provide excellent time-stratigraphic horizons in the geologic record owing to their geologically instantaneous accumulation and widespread distribution patterns. The geographic regions in which volcanic ash is an important component of marine sediments are indicated in figure 4.1. The study of widespread marine volcanic ash layers and terrestrially equivalent eruptive products has now progressed to a point at which important insights into eruptive mechanisms and distribution patterns may be gained (see, for example, Wilson and others 1978; Ninkovich and others 1978; Watkins et al. 1978; Thunell et al. 1979; Chapin and Elston 1979; Federman and Carey 1980; Self and Sparks 1981). In addition, the study of marine tephra deposits can provide information on the episodicity of volcanic activity (Kennett et al. 1977; Vallier and Kidd 1977; Hein et al. 1978), as well as on the relationship between paleoclimatic change and explosive volcanism (Kennett and Watkins 1970; Kennett and Thunell 1975 and 1977; Bray 1977 and 1979). The techniques employed in recent studies have revealed that some of the most widespread volcanic ash layers known to exist in the geologic record have occurred in the late Pleistocene. The Toba eruption (75 KYBP) is the largest known Quaternary eruption and covers the northern Indian Ocean (fig. 4.2) as a discrete ash layer over a 20,000 km2 area; the dispersed fine ash covers a much larger area (Ninkovich and others 1978). Eruptions in the Gulf of Mexico region have not produced tephra as widespread as the Toba ash. The late Pleistocene volcanic eruptions of Mexico and northern Central America, however, produced widespread tephra layers in the western Gulf of Mexico (fig. 4.3) (Kennett and Huddleston 1972; Thunell 1976; Drexler and others 1980). The Y/8 ash layer 84 KYBP is found as dispersed ash as far east as the Straits of Florida and as far south as the Carnegie Ridge (Drexler and others 1980) (fig. 4.4).

The present tephrochronology in the Gulf of Mexico was developed by establishing first the foraminiferal biostratigraphy for

This chapter appears by permission of the Colorado School of Mines, Golden, Colorado. A version of this will be published in a forthcoming work, *Subsurface Geology*, edited by LeRoy and LeRoy, 5th edition.

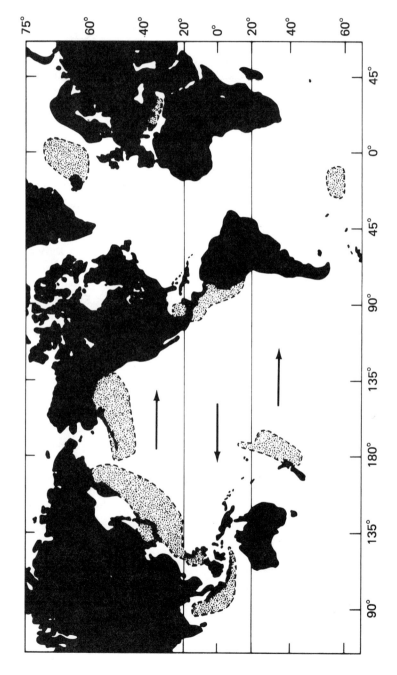

Figure 4.1 Present-day regions where ash is an important component in marine sediments. Prevailing wind patterns in the lower troposphere are indicated by arrows. (Figure prepared by D. N. Ninkovich) (From Kennett 1981. Copyright © 1981 John Wiley & Sons, Inc. Reprinted with permission of John Wiley & Sons, Inc.)

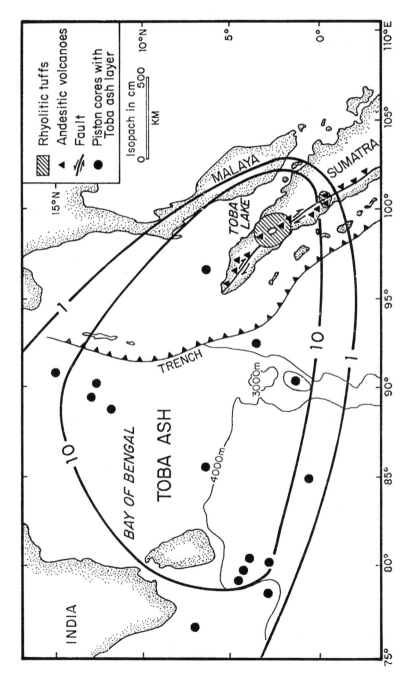

Figure 4.2 *Tephra from the Toba Eruption (75 KYBP) provides a widespread stratigraphic horizon over the northern Indian Ocean. (After Ninkovich et al. 1978.)*

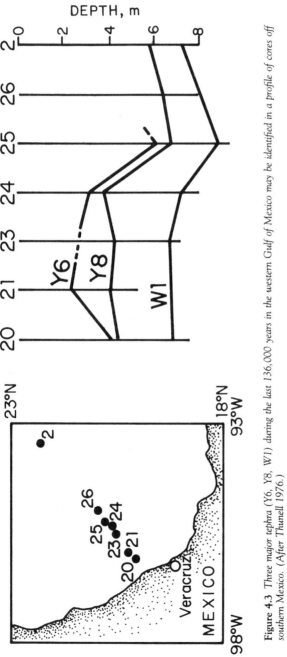

Figure 4.3 Three major tephra (Y6, Y8, W1) during the last 136,000 years in the western Gulf of Mexico may be identified in a profile of cores off southern Mexico. (After Thunell 1976.)

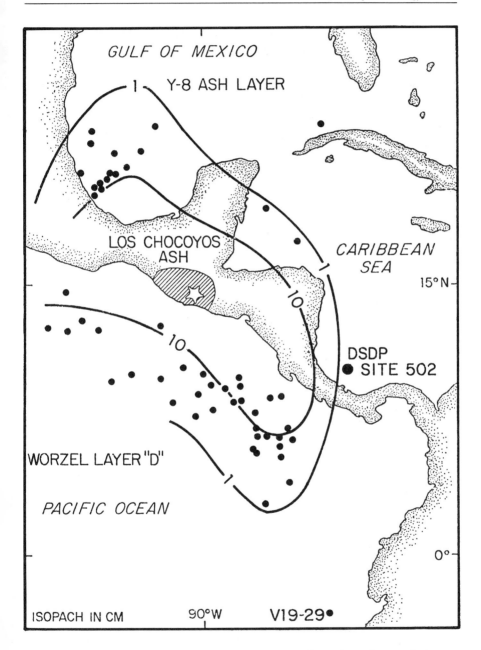

Figure 4.4 *Tephra from the Los Chocoyos eruption (85 KYBP) provides a widespread stratigraphic horizon over the eastern equatorial Pacific and western Gulf of Mexico. (After Drexler et al. 1980.)*

piston cores in the western Gulf (Kennett and Huddleston 1972; Thunell 1976). Each ash layer was assigned an age (and name) based on position within the foraminiferal zonation of Kennett and Huddleston (1972). The distribution of each ash layer was delineated only after the biostratigraphy was available for correlation of individual ash layers. This method of applying tephrochronology in the Gulf of Mexico region and elsewhere has been improved considerably.

Background

It was not until the development of the piston corer in the late 1940s (Kullenberg 1947) that the temporal nature of volcanic activity and the regional correlation of tephra layers could be examined in deep-sea sedimentary sequences. Prior to this, several studies had been successful in mapping the areal distribution of ash layers in recent deep-sea sediments from the North Atlantic (Bramelette and Bradley 1941) and the western Pacific (Keunen and Neeb 1943). The Swedish Deep-Sea Expedition (SDSE) of 1947–1948 collected a suite of sediment cores from throughout the major ocean basins that served as the first adequate data base for marine tephrochronology. Owing in part to the material collected during this expedition, the Mediterranean Sea became one of the first regions in which extensive marine tephrochronologic studies were carried out (Mellis 1948 and 1954; Norrin 1958). Other early studies clearly documented the distribution of volcanic ash layers over vast areas of the eastern Pacific and Gulf of Mexico (Worzel 1959; Ewing et al. 1959).

Since these pioneering studies, many advances have been made in the field of marine tephrochronology. The geochemical techniques used to correlate ash layers and identify source regions have become more sophisticated (Bowler 1973; Richardson and Ninkovich 1976; Watkins et al. 1978; Drexler et al. 1980). Refinement of magnetostratigraphy and oxygen isotope stratigraphy time scales have permitted more precise dating of individual ash layers (Ninkovich and Shackleton 1975; Thunell et al. 1979), and the development of new analytic techniques has made the study of fine-grained, dispersed ash possible (Huang et al. 1973 and 1975). With these advances, marine tephrochronology has become an important aspect of deep-sea stratigraphy. For a comprehensive review of previous work in marine tephrochronology, see the recent compilation by Kennett (1981).

Methods in Tephrochronology

Volcanic ash layers are distinctive lithologic units in marine or terrestrial sediments. Correlations based on lithology alone, however, can prove hazardous in areas with multiple eruptions of megascopically indistinguishable ashes. Early tephrochronologists recognized the potential problems inherent in correlating widely separated tephra, and developed techniques to separate otherwise similar tephra.

Index of Refraction of Glass Shards

The earliest successful efforts to differentiate separate tephra was based on the refractive index of volcanic glass shards (fig. 4.5) (Ninkovich and others 1964; Ninkovich and Heezen 1967). The refractive index of several glass shards from individual ash layers was determined, and layers with the same index were correlated. Although this technique has the advantage of utilizing small samples, several problems arise from this method. The initial problem is to determine the refractive index with great accuracy. The required precision makes it very important to have a well-calibrated set of refractive oils and a temperature-controlled laboratory. After obtaining a very precise, and, it is hoped, accurate index, the usual result is to find only a small difference in the refractive index of separate tephra. This is especially true for areas, such as an island arc setting, with multiple eruptions from several sources. Owing to these inherent problems, the refractive index is an unsatisfactory technique for correlation compared to the recent more sophisticated, and faster geochemical methods available.

X-ray Fluorescence of Tephra

The advent of x-ray fluorescence techniques, which can be used on relatively small samples, allowed tephrochronologists to determine a geochemical fingerprint for ash samples recovered in marine cores. Bowles and others (1973) used the method to analyze for major and minor element geochemistry of 128 volcanic ash samples from 56 cores in the eastern equatorial Pacific Ocean.

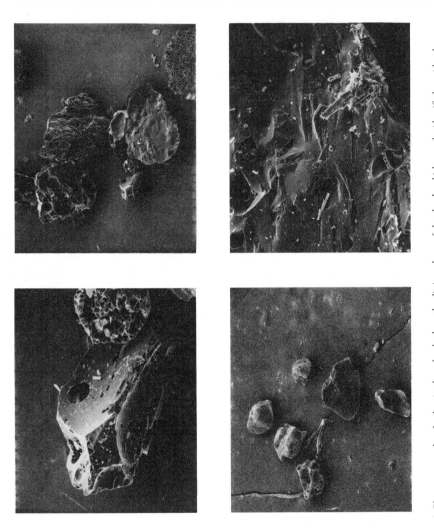

Figure 4.5 Photomicrographs of volcanic glass shards show the distinct shape of the shards which may be identified as either layers or dispersed zones in marine sedimentary sections.

In that study, Bowles and others (1973) were able to distinguish several widespread ash layers and many smaller eruptions with a limited fallout zone. One problem with the x-ray fluorescence technique is that it has been used on bulk samples, which are not restricted to glass shards. As a result, the inclusion of a few xenoliths of plagioclase, mica, hornblende, and other minerals that commonly represent 1–5% of ash layers may bias the geochemical fingerprint. If another sample of the same ash contains a different proportion of crystals or has a slightly different crystal mineralogy, the geochemical correlation of the tephra may be in jeopardy. In the Bowles and others (1973) study, many of the tephra could not be correlated, and the accuracy of the x-ray fluorescence technique may have contributed to the problem.

Microprobe Geochemistry of Tephra

The major advance in tephrochronology occurred with the application of the electron microprobe to geochemical fingerprinting of tephra. The electron microprobe may be used to focus a narrow electron beam on small (30-μm) grains and obtain a major-element analysis of the chemical composition. The eight elements analyzed in tephra research are: Si, Al, Fe, Mg, Ca, Na, K, and Ti. The elemental composition is usually expressed as oxide percents: SiO_2, Al_2O_3, FeO (total iron), MgO, CaO, Na_2O, K_2O and TiO_2. Variations in the major-element geochemistry of different eruptions from the same source of eruptions from different sources are sufficiently large to allow separation of distinctly different tephra. The advantages of the microprobe technique are the precision of the analysis (commonly percent standard deviation 1–10%), small sample size required (10–12 glass shards picked under a microscope), and the short analysis time (a few seconds to minutes for mounted samples). The technique is used exclusively by modern tephrochronologists and is now being applied to fingerprint the dispersed ash beyond the limits of megascopic ash layers, to extend stratigraphic markers over even larger areas.

The microprobe geochemical fingerprinting technique has been used to distinguish multiple volcanic ash layers in several areas beyond the Gulf of Mexico region. This is discussed in the next section. Some of the most detailed results come from the eastern Mediterranean Sea (Watkins et al. 1978; Thunell et al. 1979; Federman and Carey 1980) and serve as an example of a regional tephrochronology. The microprobe geochemical technique was used to identify six tephra horizons in widely spaced cores in the basin (fig. 4.6) for tephra deposited during the last 120,000 years (Federman and Carey 1980). The differences in geochemistry among the six tephra are readily apparent (table 4.1). Without the microprobe geochemistry, however, the correlation of the 43 ash layers found in 23 piston cores in an area of widely varying sedimentation rates would be difficult. In fact, the microprobe geochemistry of the Acrotiri Ash (Federman and Carey 1980) (fig. 4.6, table 4.1) indicated the ash was miscorrelated in one core in which the index of refraction was used for correlation (Keller et al. 1978).

Correlation of Geochemically Fingerprinted Tephra

The microprobe geochemical data for ash layers in an area of multiple volcanic sources and eruptions reveals differences between layers. These differences must be quantified, however, in order to correlate objectively discrete tephra in an area. A visual comparison of geochemical data (e.g., table 4.2) will be sufficient to detect differences, but in a complex volcanic province of multiple eruptions that are closely spaced in time, the correlations may be too subjective.

The most commonly used technique to separate individual tephra is to plot the projected weight percent FeO vs. K_2O vs. $CaO + MgO$ on a ternary diagram. The analyses of each sample of tephra are plotted, and a field is drawn to encompass all data. An example from the tephrochronology of the eastern Mediterranean (fig. 4.6, table 4.1) (Federman and Carey 1980) is shown for illustration of the method in figure 4.7. The data from each of the six ash layers (fig. 4.6) are from separate fields which delineate the range of values encountered for each tephra. It is important to observe that the fields are drawn around analyses of samples known or suspected to be correlated. This may be aided by the use of other stratigraphic information. An unknown sample may be plotted on the ternary diagram, and the correlation based on the data field intersected. If compositions of tephra overlap on the ternary diagram, then Harker diagrams, which plot every oxide (wt. %) vs. SiO_2 (wt. %), may be used to

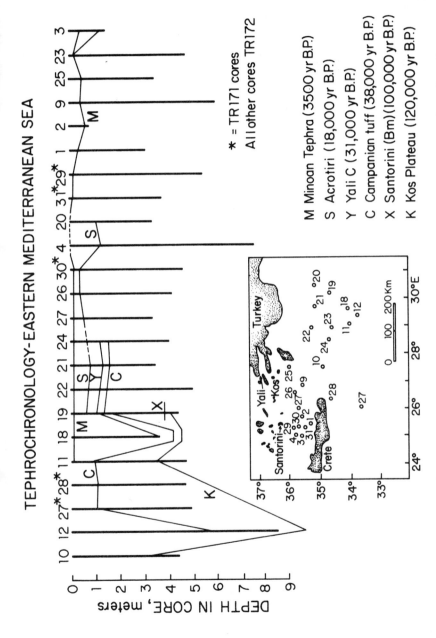

Figure 4.6 *Correlation of the six major tephra horizons in the eastern Mediterranean Sea was accomplished by microprobe geochemical "fingerprinting" of glass shards from 23 piston cores (locations shown in inset). (After Federman and Carey 1980.)*

Layer (Age)	Symbol in Fig. 5	SiO_2	Al_2O_3	FeO	MgO	CaO	Na_2O	K_2O	TiO_2	Total
Minoan (3,300 YBP)	M	72.5 (1.1)	13.7 (0.4)	1.9 (0.1)	0.26 (0.02)	1.3 (0.1)	4.9 (*)	3.1 (0.2)	0.26 (0.26)	97.7
Acrotori (18,000 YBP)	S	71.5 (1.1)	14.3 (0.7)	2.9 (0.2)	0.38 (0.04)	1.8 (0.3)	5.7 (*)	2.8 (0.1)	0.42 (0.03)	99.8
Yali C (31,000 YBP)	Y	71.8 (1.2)	14.4 (0.4)	1.8 (0.1)	0.46 (0.06)	1.8 (0.2)	4.5 (*)	3.5 (0.2)	0.32 (0.03)	98.6
Campanian (37,000 YBP)	C	61.4 (1.2)	18.3 (0.4)	2.8 (0.1)	0.35 (0.08)	1.8 (0.2)	6.9 (*)	7.2 (0.4)	0.37 (0.03)	99.1
Middle Pumice (100,000 YBP)	X	63.6 (1.6)	15.5 (0.4)	5.7 (0.4)	1.4 (0.04)	4.0 (0.5)	4.5 (*)	2.4 (0.3)	1.1 (0.1)	98.2
Kos Plateau (120,000 YBP)	K	73.5 (1.2)	12.0 (0.2)	0.42 (0.08)	0.03 (0.02)	0.5 (0.03)	4.2 (*)	4.2 (0.2)	0.05 (0.02)	94.9

Table 4.1 *Microprobe Geochemical "Fingerprints" (Standard Deviation) in wt % of Major Tephra in the Eastern Mediterranean Sea*

Source: Federman and Carey 1980.

*Not reported.

Land	(Symbols Fig. 5)	M	S	Y	C	X	K
Minoan		<u>4.1</u>	18.5	12.4	40.5	42.5	50.2
Acrotori Ignimbrite		16.2	<u>7.4</u>	22.6	26.7	37.2	104.0
Yali C		19.9	21.9	<u>5.5</u>	36.5	43.7	48.2
Yali D		38.5	49.4	40.3	66.5	64.5	19.5
Campanian Ignimbrite		37.5	47.8	33.9	<u>5.7</u>	71.7	93.6
Ischia Salina		29.0	43.0	28.6	17.2	61.7	81.1
Santorini Middle Pumice		60.6	39.6	59.8	51.9	<u>6.9</u>	133.1
Kos Plateau		49.1	59.3	48.9	87.4	72.5	<u>3.9</u>

Table 4.2 *Correlation of Terrestrial Sources and Marine Tephra in the Eastern Mediterranean Sea Using Percent Coefficient of Variation of* SiO_2, Al_2O_3, FeO, CaO, *and* K_2O

Source: Federman and Carey 1980.
Note: Lowest value indicates greatest similarity and is underlined in table.

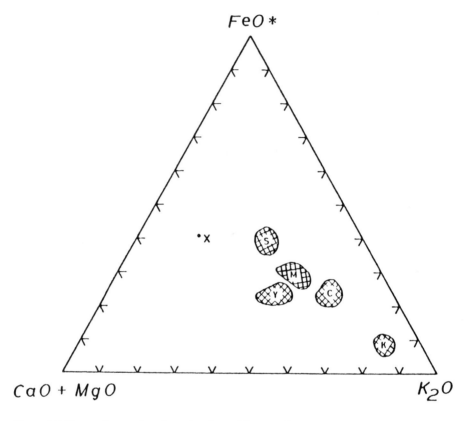

Figure 4.7 Ternary diagram of geochemical analyses of the six Mediterranean tephra correlated in figure 4.5. (After Federman and Carey 1980.)

differentiate tephra. The most commonly used plot is Al_2O_3 vs. SiO_2.

Another method for delineating separate tephra with the geochemical data is to use a similarity matrix of the data (Borchardt et al. 1971). The geochemical results for individual samples within each ash layer are averaged, and a standard deviation calculated. The correlation of samples within a layer and the between-layer correlation may be tested by calculating a percent coefficient of variation (CV) of the averaged data and standard deviation. In this method, a CV value is chosen that isolates samples of the same layer and separates different tephra. Table 4.2 is an example of the technique applied to the correlation of marine ash layers in the eastern Mediterranean Sea with the suspected terrestrial-equivalent source rocks (Federman and Carey 1980). In this example, the low percent CV between the marine ashes and six of the eight terrestrial ashes was used to establish the correlation to source.

Age Assignment in Tephrochronology
In early tephrochronology studies, the age of an ash layer was determined by developing a detailed biostratigraphy in the sedimentary core or section which contained the tephra. If more than one core or section was studied, the same procedure was repeated at every site, to be certain of the tephra correlation. As a result, the tephrochronology was not used as a true correlative tool because the detailed stratigraphy was developed at all sites in order to assign an age to each occurrence of the tephra (for example, Kennett and Huddleston 1972 and fig. 4.3). With the advent of microprobe geochemical fingerprinting however, it is possible to develop a detailed biostratigraphy and/or oxygen isotope stratigraphy in *one or two* cores from a basin with multiple tephra horizons, and extrapolate ages from this "type section" to other cores.

Excellent examples of how the method is applied may be found in the eastern equatorial Pacific Ocean and western Caribbean Sea. Two widespread volcanic ash layers were identified in sedimentary cores from the eastern equatorial Pacific Ocean (Bowles et al. 1972). The distribution of one of those layers (D-layer) was delineated on the basis of microprobe geochemistry and trace elements (Drexler et al. 1980) (fig. 4.4). The distribution of the other (L-layer) was based on the index of refraction of glass shards (Ninkovich and Shackleton 1975). The distribution of both layers has been extended beyond the limits of the previous studies with microprobe geochemistry (fig. 4.8). Both tephra were traced into piston core V19–29 (fig. 4.4; Carnegie Ridge, west of Ecuador), which contains one of the highest-resolution late Pleistocene oxygen isotope records available (Ninkovich and Shackleton 1975). Layer L is a megascopic ash layer, but Layer D is a dispersed zone of ash in V19–29 (fig. 4.9). The position of the ash layers within the oxygen isotope stratigraphy allowed very precise dating; Layer D was deposited 84 KYBP, and Layer L 230 KYBP (Ninkovich and Shackleton 1975; Drexler et al. 1980) (fig. 4.9). Therefore, the geochemically fingerprinted ash Layers D and L may be used to assign an age (84 and 230 KYBP, respectively) to the horizon where they occur in other cores in the region (fig. 4.8). When used in this manner, the tephrochronology provides time-stratigraphic information and may be used as a correlative tool in the absence of other stratigraphic data. The tephrochronology may also be used to extrapolate an absolute age into a complete or partial stratigraphic section where only a relative stratigraphy is available. This may be particularly useful in sections where the stratigraphic method applied involves pattern-recognition of fluctuations in biotic or isotopic parameters, which may be miscorrelated in the absence of other stratigraphic control.

Another example of age assignment in tephrochronology is demonstrated in the western Caribbean Sea. The tephra abundance of both megascopic and dispersed ash layers was determined in a 226m composite section (fig. 4.10) retrieved by the hydraulic piston corer at DSDP Site 502 (Ledbetter 1982) (fig. 4.4). Peaks in abundance of dispersed tephra and megascopic ash layers were numbered for reference and dated by applying biostratigraphy (fig. 1.9), magnetostratigraphy (fig. 1.8), and oxygen isotope stratigraphy (fig. 3.11) in the same core (Ledbetter 1982) (fig. 4.11). The excellent stratigraphic resolution available at DSDP Site 502 allows establishment of a complete tephrochronology for the western Caribbean. Unfortunately, however, the western Caribbean did not receive tephra from most of the major eruptions in the area (for example, see fig. 4.4). For that reason, the late Pleistocene tephrochronology of the Gulf of Mexico was developed independently.

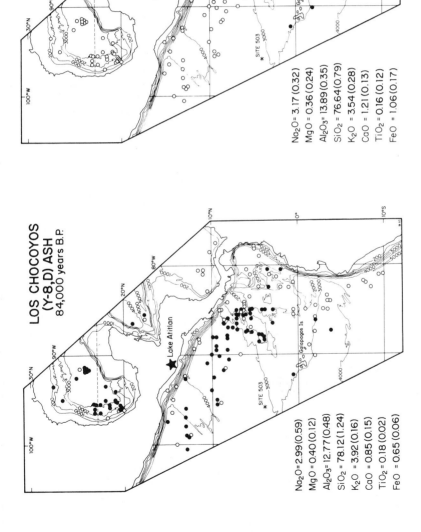

Figure 4.8 *The distribution of two major ash layers (solid circles) in the eastern equatorial Pacific Ocean was determined by geochemical analysis of ash layers found in cores (circles) from the region.*

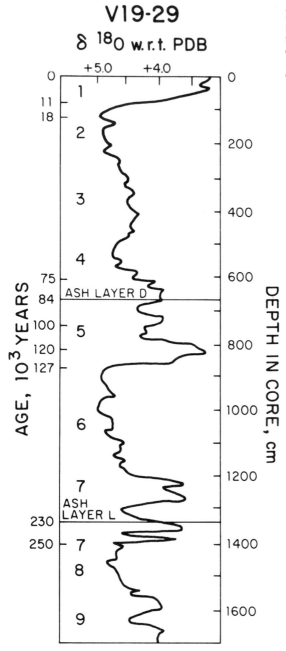

Figure 4.9 *The age of Ash Layer D (Los Chocoyos Ash) and Ash Layer L may be determined by geochemically correlating the tephra into core V19–29 (see fig. 4.3) with high-resolution oxygen isotope stratigraphy.*

SITE 502 TEPHRA ABUNDANCE

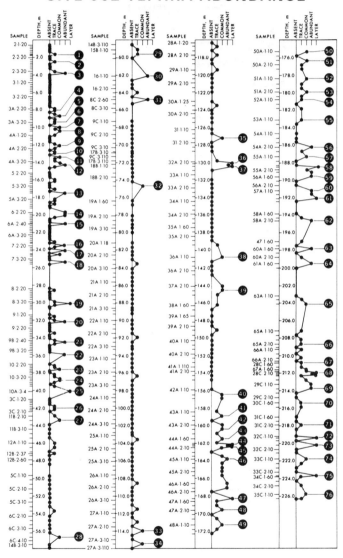

Figure 4.10 *Volcanic ash abundance in the 226m composite section at Site 502 (see fig. 4.3). Abundance peaks are numbered for reference to figure 4.10.*

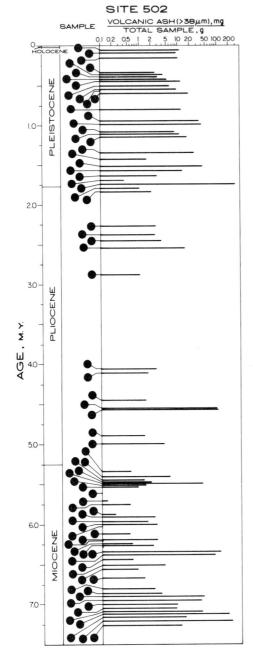

Figure 4.11 *The recovery of a long, complete section at Site 502 (see figs. 4.3, 4.9) with the hydraulic piston corer on board the "Glomar Challenger" provides an opportunity to develop a regional tephrochronology for several million years.*

Late Pleistocene Tephrochronology in Gulf of Mexico Region

Late Pleistocene volcanic eruptions in the trans-Mexican volcanic belt in southern Mexico area (for example, Mahood 1980; Wright 1981) and in Guatemala (Rose et al. 1979; Hahn et al. 1979; Drexler et al. 1980) have produced widespread blankets of pumice, ignimibrites, and airfall ashes in the region adjacent to the western Gulf of Mexico. The volume of eruptive products on land suggests that marine tephra layers from these eruptions are to be expected in the ocean basins surrounding southern Mexico and northern Central America. Only one of the terrestrial ashflows, however, has been correlated to a marine ash layer; the Los Chocoyos ash from the Lake Atitlan caldera in Guatemala was correlated with the D-Layer ash in the Pacific Ocean and the Y/8 ash in the western Gulf of Mexico (Hahn et al. 1979; Drexler et al. 1980) (fig. 4.4). Correlation of other major eruptions to marine tephra has not been accomplished to date, although work is continuing.

In spite of the lack of correlation with terrestrial volcanic sources, 11 marine tephra in the Gulf of Mexico region have been correlated geochemically (figs. 4.8 and 4.12). Only three of those tephra (Y/6, Y/8, W/1) have been identified as widespread tephra horizons in the Gulf of Mexico (figs. 4.8 and 4.12). Of those three, the Y/8 ash (Kennett and Huddleston 1972) is the most widespread in the western Gulf of Mexico. The Y/6 and W/1 tephra are clearly less widespread than the Y/8 (Los Chocoyos) ash because they are not found in the Pacific Ocean and do not extend as far east as the Y/8 tephra (fig. 4.11). Not shown in fig. 4.12 are three other late Pleistocene tephra that are found in the western Gulf of Mexico (Rabek, 1983). The areal distribution of those layers is so limited that they occur only in a few cores immediately adjacent to Mexico. Those layers are found in the W/2, X/5, Y/5 foraminiferal substages in the zonation of Kennett and Huddlestun (1972). The W/2 ash may be more widespread than the current limited distribution would suggest, however, because this tephra horizon is found in only a few long piston cores that penetrate sediment of the appropriate age (150,000 yr.).

Dating of the Gulf of Mexico tephra was refined from the original tephrochronology established by Kennett and Huddles-

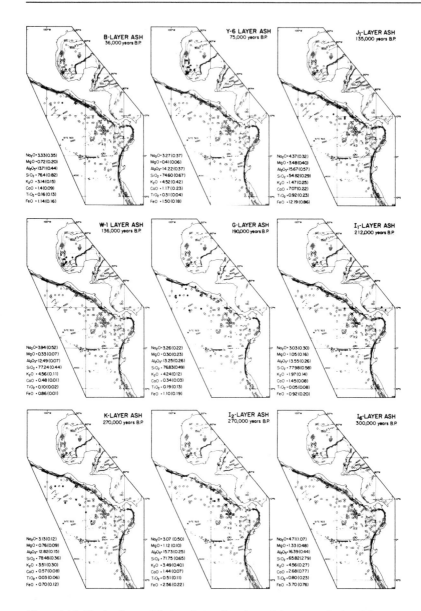

Figure 4.12 *The distribution* (filled circles) *and geochemical fingerprint of eight ash layers* *(names from Bowles et al. 1973) in the Gulf of Mexico region.*

ton (1972). In that study, a subzonation of the *Globorotalia menardii* stratigraphy (Ericson and Wollin 1968) was developed, and the correlation of volcanic ash layers was based on the biostratigraphy. As a result, the name of each tephra was based on the foraminiferal subzone (fig. 4.3) (Kennett and Huddleston 1972). More recently, the geochemical fingerprinting method has proved ideal for dating tephra. For example, both the Y/8 and W/1 ash layers were dated by tracing the geochemically fingerprinted tephra (table 4.3) into cores with a high-resolution oxygen isotope stratigraphy. The Y/8 ash was correlated geochemically to the D-Layer in the Pacific and dated by tracing the fingerprint to a dispersed-ash zone in piston core V19–29. The oxygen isotope stratigraphy in core V19–29 (Ninkovich and Shackleton 1975) provides a detailed stratigraphy, and the Y/8 ash was dated at 84 KYBP (fig. 4.9) (Drexler et al. 1980). The W/1 ash was dated at 136 KYBP by tracing the geochemically fingerprinted tephra into an oxygen isotope stratigraphy (Falls 1980) in a piston core from the western Gulf of Mexico (fig. 4.13).

The late Pleistocene tephrochronology of the western Gulf of Mexico reveals three widespread, and three more limited, tephra horizons during the last 140,000 yr. The frequency of tephra horizons, therefore, is 1 per 23,000 yr. If the entire Pleistocene section were available, such a frequency would yield over 70 volcanic ash layers during the Pleistocene. If available, a Pleistocene tephrochronology would provide a detailed, easily obtainable stratigraphic tool in the basin.

Applications of Tephrochronology in the Gulf of Mexico Region

The geochemically fingerprinted tephra horizons offer an easy method for establishing a stratigraphy in newly recovered marine and terrestrial sections. Instead of acquiring a more time-consuming biostratigraphy in every new core in a basin, the tephrochronology may be used for an initial stratigraphy. More important, perhaps, the tephrochronology may be used to correlate another stratigraphy to the time scale. For example, a stratigraphic method that depends on correlating a pattern of fluctuations in values to a standard pattern may be correlated incorrectly, owing to undetected hiatuses or changes in sedimentation

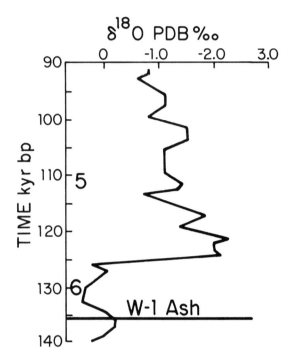

Figure 4.13 *The oxygen isotope stratigraphy of a core in the western Gulf of Mexico was used to date the W/1 ash (see figs. 4.2 and 4.11). (After Falls 1980.)*

Layer (Age)	SiO_2	Al_2O_3	FeO	MgO	CaO	Na_2O	K_2O	TiO_2
Y–6 (75,000 YBP)	74.60 (0.67)	14.23 (0.37)	1.50 (0.18)	0.41 (0.06)	1.17 (0.23)	3.27 (0.37)	4.52 (0.42)	0.31 (0.04)
Y–8, "D" (84,000 YBP)	78.12 (1.24)	12.77 (0.48)	0.65 (0.06)	0.40 (0.12)	0.85 (0.15)	2.99 (0.59)	3.92 (0.16)	0.18 (0.02)
W–1 (136,000 YBP)	77.24 (0.44)	12.49 (0.07)	0.86 (0.01)	0.33 (0.07)	0.48 (0.01)	3.94 (0.52)	4.56 (0.11)	0.10 (0.02)

Table 4.3 Microprobe Geochemical "Fingerprints" (Standard Deviation) in wt % of Three Major Tephra in the Gulf of Mexico (calculated to 100%)

rate. To alleviate the problem, the tephrochronology of one or more ash layers may be used to confirm the age of the studied section and constrain the correlation of the fluctuations to the time scale.

The maximum extent of tephra layers in the Gulf of Mexico (figs. 4.3; 4.12) is limited to the western part of the Basin. The tephrochronology may be extended, however, throughout the Basin because a dispersed zone of tephra may be identified beyond the area of megascopically identifiable ash layers (fig. 4.14). The tephra that fell in this zone was dispersed among the dominant lithology due to the lower volume of material reaching the far downwind location. The decreased volume may have been responsible directly for the dispersed nature of the ash or indirectly responsible because any thin layers formed originally in this zone would be dispersed by bioturbation. The Caribbean tephrochronology shown in figures 4.10 and 4.11 included dispersed tephra horizons where tephra less than 38 μm in size was detected by routine sedimentological or micropaleontological examination of the coarse fraction.

Therefore, with the inclusion of both dispersed and megascopic tephra horizons, the tephrochronology may be extended to include the entire Gulf of Mexico. The Y/8 dispersed-ash zone has been found as far east as the Straits of Florida (fig. 4.4) (Drexler et al. 1980), and the other layers are expected to show a similar distribution pattern (fig. 4.14). Therefore, it is possible to extend a well-dated tephrochronology into marine or terrestrial sections within this region to facilitate correlation to a time scale.

Conclusions

The advent of the electron-microprobe geochemical fingerprinting technique in tephrochronology has made it possible to establish a tephra type-section for a sedimentary basin. Volcanic ash layers dated in the type-section may be correlated to undated sedimentary sections by a geochemical analysis of either *megascopic* or *dispersed* ash found in the section under study. The method provides fast, accurate ages in widely separated cores in an area the size of the Gulf of Mexico.

Tephrochronology in the Gulf of Mexico reveals at least three

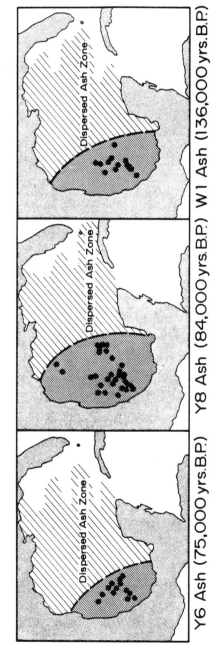

Y6 Ash (75,000 yrs.B.P.)   Y8 Ash (84,000 yrs.B.P.)   W1 Ash (136,000 yrs.B.P.)

Figure 4.14 *The area distribution of the three major tephra during the last 136,000 years in the western Gulf of Mexico (see fig. 4.2) may be extended into the eastern Gulf by identifying the dispersed ash component at more distant sites.*

major and three minor ash layers in the late Pleistocene (0–140 KYBP). Two of those layers were traced into cores with a high-resolution oxygen isotope stratigraphy, and the age assigned from that stratigraphy may be extrapolated to other sections in which the same tephra is identified geochemically.

The late Pleistocene tephrochronology in the Gulf of Mexico may be extended to include the entire Pleistocene when sections become available for study. In that event, the tephrochronology will prove to be a valuable contribution in the Pleistocene stratigraphy of the region.

5

Continental Glacial Stratigraphy, Marine Evidence of Glaciation, and Insights into Continental-Marine Correlations

Richard H. Fillon
Belle W. Baruch Institute
University of South Carolina
Columbia, South Carolina   29208

5

Introduction

The late Cenozoic glacial record is clearly distinguished by the tremendous episodic growth and subsequent decay of glaciers on continental areas of the Northern Hemisphere. The parallel academic disciplines of glacial geology and late Cenozoic paleoceanography are evidence that those large-scale environmental changes and the synergistic question of what caused the ice ages are compelling subjects to many researchers.

In continental and marine glacial studies, as in all of geology, the ability to correlate one lithology or one geologically or paleontologically important event from one location to the next, thus establishing a stratigraphy, is critical. It has been the development of the stratigraphic techniques outlined in chapters 1–4 of this volume that has been most responsible for increasing our understanding of the scope and magnitude of variations in late Cenozoic marine environment. Although not directly addressed in those chapters, the development of the paleomagnetic time scale, through radiometric dating of terrestrial volcanic rocks, and improvements in tephrostratigraphic techniques have also been pivotal to recent advances in placing terrestrial glacial events within a reliable chronostratigraphic framework (Cooke 1972; Berggren and Van Couvering 1974; among others).

The first part of this chapter is a review and discussion of the status of late Cenozoic glacial studies. The subject is addressed from both continental and marine perspectives. It starts with the classical North American continental glacial stratigraphy and, to this basic framework, adds the glacial stratigraphies of the Rocky Mountains, the Sierra Nevada, the Alps, and the northern European lowland using the best available chronostratigraphic control to construct a composite correlation with the marine record.

The last part of the chapter is devoted to an analysis of the basis of continental-marine correlation in the late Cenozoic. Emphasis is placed on the likely control of terrestrial and marine climate by a single driving mechanism. That mechanism is the systematic alteration in the distribution of solar energy over the Earth caused by periodic and, therefore, predictable changes in the Earth's orbital configuration. Orbital control of climate and glaciation was first proposed by Milankovitch in 1941 and has recently been the focus of much research. Indications are that the theory of orbital climatic control will eventually unify studies of the late Cenozoic much as the theory of seafloor spreading unified the study of older parts of the geologic record. The chapter concludes by presenting a comparison of orbitally predicted paleoclimatic cycles with the evolving chronology of late Cenozoic continental and marine events.

North American Glacial Stratigraphy of the Midcontinent

Introduction
After Louis Agassiz (1840) published his idea of a European ice sheet in *Etudes sur les Glaciers*, American geologists rapidly adopted the concept that large portions of North America had once been glaciated. Thirty-five years later, McGee (1878) was probably the first American geologist to recognize evidence for an interglacial interval in the presence of an organic tree-bearing horizon separating two tills in eastern Iowa. The idea of multiple North American glaciations then gained increasing acceptance. By 1909, with Shimek's work in Nebraska, the commonly accepted modern system of Nebraskan, Aftonian, Kansan, Yarmouthian, Illinoian, Sangamonian, and Wisconsinan stages had evolved (fig. 5.1). A fifth glacial stage, the Iowan, was named by Chamberlin (1895; 1896) and fitted between the Illinoian and the Wisconsinan. Its significance was hotly debated (Ruhe and Scholtes 1959; Leighton and Brophy 1966; and Leighton 1968), but the Iowan glacial stage was finally rejected as a substage of the Wisconsinan (Frye and Willman 1960).

By definition the Iowan postdated the Sangamon, and therefore, deposits assigned to it probably lay within the early Wisconsinan, presumably within the Altonian glacial substage of Frye and Willman (1960). The early confusion surrounding the Iowan reflects the accommodations that often have to be made in fitting local events into a continental framework. Examination of the Wisconsinan stratigraphic scheme illustrated by Willman and Frye (1970) (fig. 5.2) indicates three glacial advances into Illinois from Wisconsin during Altonian (substage) time and two soil forming interstadial episodes. One of the three Altonian glacial

Figure 5.1 *The status of middle North American glacial stratigraphy in the first decade of the twentieth century.*

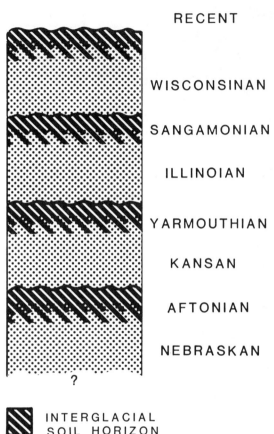

RECENT

WISCONSINAN

SANGAMONIAN

ILLINOIAN

YARMOUTHIAN

KANSAN

AFTONIAN

NEBRASKAN

?

 INTERGLACIAL SOIL HORIZON

 GLACIAL TILL

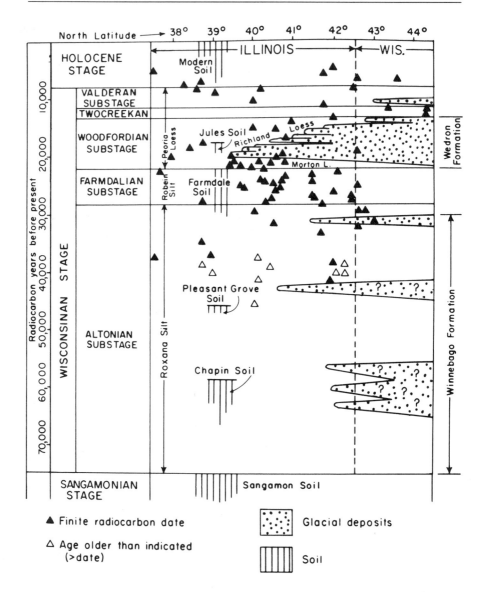

Figure 5.2 *Time-distance diagram showing glacial deposits, soil horizons and radiocarbon dates in the Wisconsinan and Holocene Stages in Illinois. (From Willman and Frye 1970. Reprinted by permission of the Illinois State Geological Survey.)*

events could well have been related to the type Iowan in eastern Iowa, but confirmation of this will have to await a considerable improvement in radiometric dating techniques.

The Wisconsinan Stage

From a lithostratigraphic-morphostratigraphic viewpoint, the Iowan was just not sufficiently prominent or widespread to facilitate correlation on a continental scale. The younger Woodfordian (late Wisconsinan) is prominent and widespread as well as being in the range of accurate ^{14}C dating (to about 40 KYBP) and so has been widely correlated. In a sense, the Woodfordian is the "backbone" of the Wisconsinan because it provides a locus to which to relate all other post-Sangamonian glacial deposits. The Woodfordian was so widespread that it buried or obliterated many of the earlier glacial deposits and thereby enhanced its preeminence. This should not be construed as implying that Altonian glacial events were minor; after all, the extension of glacier ice into Illinois implies a large volume of continental ice on North America.

The presence of several well-developed soil profiles (Chapin, Pleasant Grove, and Farmdale) between the early Altonian and the Woodfordian further suggests several significant retreats of the ice within the early–mid-Wisconsinan. Just how far Altonian ice retreated during those soil-forming episodes is largely a matter of conjecture at the present time, but there is some evidence that Hudson Bay may have been deglaciated several times during the Wisconsinan (Shilts et al. 1981; Andrews et al. 1983) (fig. 5.3).

Andrews et al. (1983) measured amino acid ratios of mollusk shells in Bell Sea deposits near Hudson Bay. On the basis of mollusk and pollen assemblages, these shells are thought to be representative of interglacial conditions. Next, Andrews and colleagues determined the amino acid ratios for numerous shell fragments from Wisconsinan-age tills around Hudson Bay. They found three distinct sets of ratios between Bell Sea interglacial age and Tyrell Sea postglacial age. If the age of the Bell Sea interglacial is considered to be about 120 KYBP, which is coincident with the last "interglacial" of the marine isotope record (substage 5e) (Emiliani 1966; Shackleton and Opdyke 1973; see also chap. 3) their results would imply that the North American ice sheet was sufficiently reduced in volume during Wisconsinan interstades to allow entry of marine waters into Hudson Bay.

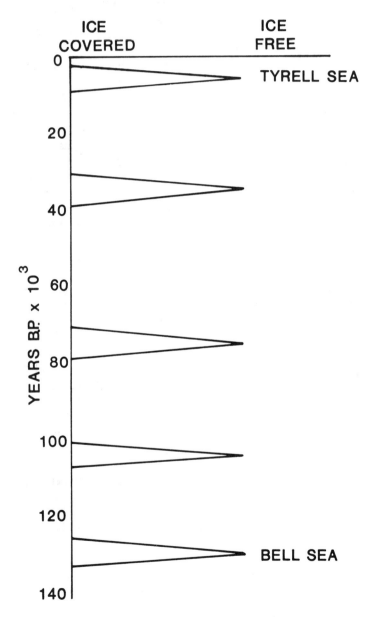

Figure 5.3 *Chronology of deglaciation of Hudson Bay during the Wisconsinan based on amino acid racemization studies (after Andrews et al. 1983).*

The Sangamonian Stage

Although glacial stages are defined on the basis of till deposits which defy long-distance correlation, interglacial stages are based on the presence of prominent more widely correlatable soil profiles or paleosols. At a few localities, mollusc and vertebrate fossils have been found associated clearly with Sangamon soils. The nature of the faunas suggests a moderate climate, perhaps even warmer than the present (Hibbard et al. 1965; Taylor 1965).

If Bell Sea "interglacial" sediments are of Sangamonian age, correlations between Altonian soils (fig. 5.2), and episodes of deglaciation in Hudson Bay might be possible. However, there are as yet no radiometric dates to support direct correlation of the Bell Sea deposits with isotope substage 5e, and moreover, no direct correlation between substage 5e and the Sangamon. Substage 5e could just as well correlate with either of the pre–^{14}C-range Altonian soils.

Fossils, when present, provide a means of distinguishing one paleosol from another (see, for example, Hibbard et al. 1965). The presence of radiometrically datable geochemically distinctive volcanic ash layers in systematic association with soil horizons, or the absence of such ash layers also provides a means of differentiating paleosols (see, for example, Reed et al. 1965). Of the three midcontinent interglacial soils, only the Sangamon is not associated with distinctive volcanic ash beds.

The Illinoian Stage

Although it was recognized in 1897 that the Wisconsinan required division into several substages (Leverett 1897), the pre-Sangamonian, Illinoian Stage remained enigmatic until 1950, when a useful subdivision was developed (Leighton and Willman 1950). Today three Illinoian substages are recognized, the Liman, Monican, and Jubileean (fig. 5.4), representing three distinct glacial advances and several soil-forming episodes in Illinois (Willman and Frye 1970). The glaciation of the earliest substage (the Liman) was the most extensive, whereas the glaciation of the latest substage (the Jubileean) was the least extensive. The Liman advance would have erased much of the evidence of any earlier, but less extensive, Illinoian glacial advances. Therefore, even in regions of minimal disruption by Wisconsinan ice, it cannot be taken as a certainty that there was only one impor-

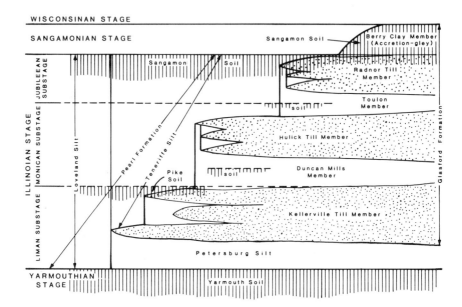

Figure 5.4 *Schematic cross-section showing the relations of Illinoian age glacial tills, soil horizons and silt deposits in western Illinois. (From Willman and Frye 1970, reprinted with permission of the Illinois State Geological Survey).*

tant glacial event between the exceptionally well-developed
Yarmouthian interglacial soil, which defines the base of the Illinoian, and the end of the Liman advance.

The Yarmouthian Stage
The pre-Illinoian Yarmouth soil is well expressed generally as a
thick interglacial weathering zone on Kansan deposits. West of
the Mississippi River the Yarmouth soil is characteristically associated with the type "O" Pearlette ash (Izett et al. 1971). The
Yarmouthian interglacial may have lasted several times longer
than the Sangamonian as indicated by the much greater thickness of Yarmouthian paleosols (Willman and Frye 1970). Yarmouthian paleosols (and underlying deposits) have been more
intensively eroded than younger glacial-interglacial strata making
long distance correlations somewhat more difficult.

The Kansan Stage
To date, the Kansan glacial stage has not been formally subdivided into substages although, as with the Illinoian, careful and
intensive field efforts may yield several distinct glacial events of
substage rank. In fact, Frye et al. (1965) suggest the presence of
two distinct tills of Kansan age in Illinois (fig. 5.5). The Kansan
Stage, being older than the Illinoian, is more poorly preserved
and less widely distributed. The Kansan is bounded below by the
Aftonian interglacial soil which is much less widespread than the
Yarmouth soil.

The Aftonian Stage
Although pre-Yarmouthian paleosols with underlying glacial deposits have been identified, those soil profiles have been assigned
to the Aftonian interglacial. In fact, however, the Aftonian type
section does not contain a soil profile (Willman and Frye 1970).
The formal Afton Soil soil stratigraphic unit was described beneath Kansan Till at an alternate site (Willman and Frye 1970).
Thus there is really no way of knowing whether the "Afton" soil
at that location is immediately underlain by glacial deposits that
are precisely equivalent to the type Nebraskan of Shimek (1909)
as is implied in figure 5.1. To illustrate the difficulties of correlating pre-Yarmouthian soils and glacial deposits, Izett et al. (1971)
point out that type "S" Pearlette ash has been found in sediments

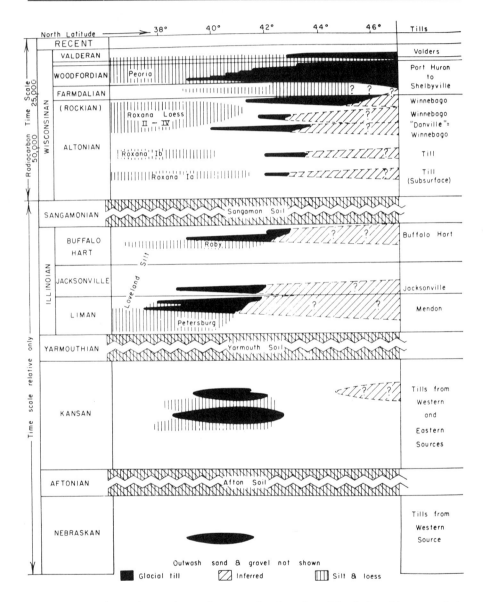

Figure 5.5 *Time-distance diagram showing the stratigraphic relationships of glacial tills, soil horizons and silt deposits in Illinois and Wisconsin. (From Frye et al. 1965, fig. 3, p. 46, and fig. 2, p. 744. Copyright ©* 1965 *by Princeton University Press. Reprinted by permission of Princeton University Press.)*

assigned both a late Kansan age (the Sappa Formation) and a late-Nebraskan age (the Fullerton Formation) in Nebraska (Reed et al. 1965). Thus the Aftonian may either postdate or predate the type "S" Pearlette ash. These two possibilities are discussed further in the section below on Midcontinental Ash Chronology.

The Nebraskan Stage

The Nebraskan, like the Kansan, has not yet been formally subdivided, although Reed and Dreezen (1964) recognize two advances of the ice during Nebraskan time. The upper limit of the Nebraskan Stage is the problematical Aftonian soil, but its lower limit was not even defined by Shimek (1909). By common usage among glacial geologists, the Nebraskan has been equated with the oldest pre-Aftonian till in the midcontinent region. This definition allows the inclusion within the Nebraskan stage of more than the one pre-Aftonian till viewed by Shimek. Recently, Boellstorff (1978) reported the results of drilling in southern Iowa that encountered two tills below the Aftonian soil. He stated (p. 305) that the lowermost till is "Pre-Nebraskan by definition" because it directly underlies the "classic Nebraskan-Aftonian-Kansan sequence." This is clearly a misstatement, because the base of the Nebraskan was never formally defined. Because Boellstorff's lower till is pre-Aftonian and obviously glaciogenic, it can only be assigned at present to the Nebraskan. It may correlate with the early Nebraskan Elk Creek Till of Nebraska (Reed et al. 1965). The lower till and the overlying paleosol and clay bed described by Boellstorff (1978) that separate the lower till from the upper Nebraskan Till may well reflect events that are only of substage rank. No pressing need therefore exists at this time to formally erect a new glacial stage based on Boellstorff's data. It should be remembered that one of the principal criteria for defining interglacials of stage rank has been the *widespread* nature of the interglacial soil.

Midcontinental Ash Chronology

Volcanic ash, including age equivalents of the type "O" Pearlette ash, the Bishop ash, the type "S" Pearlette ash and the type "B" Pearlette ash are intercalated with midcontinent glacial and interglacial deposits. Fission track dates on the ash beds of 0.6, 0.7, 1.2 and 2.2 MYBP, respectively, (Naeser et al. 1973; Boellstorff

1978), are similar to K-Ar dates for tuffaceous equivalents at localities in the western United States (Christiansen et al. 1968; Richmond 1970). The type "S" Pearlette ash (ca. 1.2 MYBP) was found originally in the lower part of the Sappa (loess) Formation. The type "S" ash is paleomagnetically reversed (Izett et al. 1971). The pre-Yarmouthian Sappa Formation is considered by Reed et al. (1965) to be a periglacial equivalent of the latest Kansan, Clarkson Till. Clarkson Till may therefore overlie the type "S" ash in certain areas, thus explaining the observed occurrence of type "S" Pearlette ash beneath, as well as above, Kansan Till (Izett et al. 1971). The 0.6 and 0.7 MYBP type "O" Pearlette and Bishop equivalent ash beds are normally magnetized and consistently overlie Kansan Till (Izett et al. 1971). They are therefore most probably Yarmouthian. However, as there is still considerable confusion surrounding stratigraphic studies that fail to differentiate the three Pearlette-type ash beds, two alternate stratigraphies, one with an expanded Kansan and one with an expanded Nebraskan are suggested (with and without brackets in figure 5.6, column 7), although a reduced Kansan may more easily explain the majority of observations.

According to Boellstorff the 2.2 MYBP ash, which may be the type "B" Pearlette ash K-Ar–dated at about 2.0 MYBP (Naeser et al. 1973), is included in a nonglacial interval. The 2.2 MYBP ash therefore dates either the Aftonian interglacial or a "mid"-Nebraskan interstade. Boellstorff's concept that the till underlying the ash is pre-Nebraskan is not adopted here.

Chronology of Glaciation Based on Dated Events in the Western Mountains

Introduction

In the midcontinent, tills and soils are spread out in layer-cake fashion, with the oldest at the bottom of the section. However, the stratigraphy of glacial deposits in the mountainous regions of western North America is quite different (Richmond 1965 and Wahrhaftig and Birman 1965). In the western mountains, the Rockies and the Sierra Nevada, the youngest tills and soils are found at low elevation in the valleys, whereas remnants of older tills and soils are found perched on the heights. Fortunately, glacial geologists working in the mountains are often aided by correlatable step-like river terraces and valley moraines that permit the assignment of relative ages and by occasional extrusive volcanic rocks and pyroclastics that can be radiometrically dated. Such radiometric dates have pushed the chronology of temperate glacial events in North America back to about 3.0 MYBP (Curry 1966).

The availability of radiometric dates on glacial deposits in the mountains has long been recognized as providing the possibility of developing a comparative chronology for the glacial tills and soils of the midcontinent. This challenge stimulated a number of recent studies that have sought to correlate the record of mountain glaciations with continental glacial events in the lowlands and thus to put North American continental glacial stratigraphy into a K-Ar time framework (Richmond 1965, 1970; Cooke, 1972; Berggren and Van Couvering 1974).

Correlation with the Midcontinent

Rocky Mountain glacial stratigraphy (fig. 5.6, column 8) is supported by K-Ar dates on volcanics intercalated with tills and soils in the Yellowstone region (Richmond 1970). Early and Middle Pinedale glacial deposits have been equated with the Woodfordian of the midcontinent on the basis of similar ^{14}C dates. The much older Cedar Ridge glaciation has been correlated by Richmond (1970) with the Kansan of the midcontinent based on a K-Ar–dated volcanic ash horizon within the upper part of the Cedar Ridge deposits, which is petrographically and chronologically equivalent to the type "O" Pearlette ash of Kansas and Nebraska. Richmond (1970) also considers that the Washakie Point glaciation predates deposition of the type "S" Pearlette ash.

Sierra Nevada glacial stratigraphy has been summarized by Wahrhaftig and Birman (1965) and Birkeland et al. (1970) (fig. 5.6, column 9). Correlation of the Tahoe, Tioga and Tenaya with the Rocky Mountain, Pinedale and correlation of the Sherwin with the Cedar Ridge are constrained radiometrically (Birkeland et al. 1970). Chronological control in the Sierra Nevada includes K-Ar dates on basalts underlying Tioga-Tenaya-Tahoe Till and bracketing the Casa Diablo Till and K-Ar dates on the Bishop tuff which caps the Sherwin Till and on basalt underlying

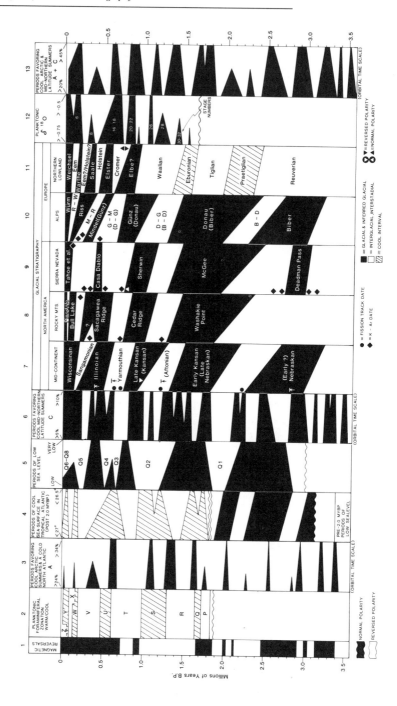

Figure 5.6 *Suggested chronology of glacial events for the last 3.5 million years. Explanation and references: Column 1—after Mankinen and Dalrymple 1979. Column 2—after Ericson and Wollin 1968 and van Donk 1976; Globorotalia menardii absent in shaded intervals. Column 3—after Fillon and Williams 1983; see also fig. 5.18; record of predicted above average (A > 24%) and maximum intensity (A > 34%) coolings. Column 4—(upper part) after Briskin and Berggren 1975; cool (< 27°C) and very cool (< 26.5°C) tropical sea surface temperatures inferred from planktonic foraminifera.*

Note: the resolution, i.e., the cut-off values in columns 3 and 4 were chosen to maximize correlation with column 2.) Column 4—(lower part) after Akers and Holk 1957 and Poag 1972; relatively low sea level (shaded). Column 5—transgressions Q1 to Q8 and regressions (shaded) after Beard et al. 1982; see also figures 5.13 and 5.22. Column 6—after Fillon and Williams 1983; see also fig. 5.19. The selection of predicted low amplitude cool periods (C > 5% and C > 10%) maximizes correlation with the continental record of glaciations (columns 7–11). Column 7—The chronology is provided by fission track ages of 0.6, 0.7, 1.2 and 2.2 MYBP from volcanic ashes (Boelstorff 1978) correlated respectively to the type "O" Pearlette ash, the Bishop ash, the type "S" Pearlette ash, and the type "B" Pearlette ash. The arrows refer to the estimated time of initiation of regressive episodes recorded in the Gulf Coast area (Akers and Holk 1957) that are correlated to planktonic foraminiferal datums (Poag 1972). Column 8—Stratigraphy after Richmond (1970) and Birkeland et al. (1971). Chronological control consists of K-Ar dates on pumice and welded tuffs from the Yellowstone region. These volcanics include beds that are correlative with the various Pearlette ash beds and the Bishop ash bed of the midcontinent. Column 9—Stratigraphy based on that of Wahrhaftig and Birman (1965). Chronological control is based on K-Ar dates on basalts and paleomagnetic measurements (Cox et al. 1965; Curry 1966; Birkeland et al. 1971). Column 10—Chronological control consists of K-Ar dates on Laacher basalts from the Rhine terraces in Germany (Richmond 1970) and from correlation of Alpine terrace deposits with paleomagnetically dated loess records (Kukla 1977). The alternative chronology of Brunnacker et al. (1982) is indicated in brackets. Column 11—Chronology after those of Cooke (1972), Berggren and Van Couvering (1974) and Kukla (1977). Evidence of a Brunhes-Matuyama boundary age for the Cromerian is cited by Kukla (1977). Pollen zone ages (Reuverian through Wadian) are as reported by Berggren and Van Couvering (1974) with the exception that the Praetiglian is shortened in agreement with van der Hammen et al. (1971) and Brunnacker et al. (1982). Column 12—after Fillon and Williams (1983); see also fig. 5.21. Isotopically heavy (> −0.75 ‰) and very heavy (> −0.5 ‰) intervals in the isotopic record of HPC Site 502B reported by Prell (1983). Column 13—after Fillon and Williams (1983); see also figures 5.20, 5.21. The selection of predicted cool (A + C > 20%) and very cool (A + C > 45%) intervals maximizes correlation with column 12.

McGee Till (Wahrhaftig and Birman 1965; Birkeland et al. 1977). Till just below the Bishop tuff is normally magnetized (Cox et al. 1965) and therefore must be representative of the basal Brunhes or possibly Jaramillo magnetozones. The Deadman Pass Tillite of the Sierra Nevada (Curry 1966) is overlain by basalt dated by K-Ar at 2.7 MYBP and underlain by another basalt dated at 3.1 MYBP. The Deadman Pass Tillite is, therefore, definitely older than the Rocky Mountain McGee Till. The McGee and Deadman Pass glaciations might thus constitute equivalents of the two Nebraskan glacial events of Reed and Dreeszen (1964) and Boellstorff (1978).

Reed et al. (1965, p. 195) stated that in Nebraska "the early-Nebraskan deposits were significantly eroded before they were covered by the late-Nebraskan deposits, developing a greater relief than is typical of the interstadial contacts of the later glaciations". The possibility of a long nonglacial interval of stage rank (at the latitude of Nebraska) between about 2.7 and 2.2 MYBP cannot therefore be excluded (fig. 5.6). This erosional mid-Nebraskan interval may explain the absence of any notable soil horizon remaining on "early" Nebraskan Till; the presence of such a soil horizon would perhaps have supported the creation of a new glacial stage (Elk Creekian) to replace "early" Nebraskan.

Pre–Deadman Pass Glacials in North America?

To date, the oldest known glacial deposit in temperate North America remains the Deadman Pass Tillite (3.1 to 2.7 MYBP, fig. 5.6). Glacial advances of greater antiquity may have occurred, but these will be very difficult to identify, owing to the incompleteness of the record. In the Sierra Nevada, where volcanic rocks provide the best materials for dating old glacial deposits, there is an unfortunate lack of such rocks with dates between about 3.2 and 7.4 MYBP. (Wahrhaftig and Birman 1965; Cox et al. 1965). This lack of suitable rocks, combined with the significant amount of uplift and erosion that has characterized the Sierra Nevada during the Neogene, effectively rules out the imminent discovery of a pre-Deadman Pass glacial deposit in this region.

In the midcontinent, dating opportunities are very limited. The 2.2 MYBP volcanic ash (Boellstorff 1978) provides the oldest radiometric date so far obtained from a glacial section east of the Rocky Mountains. Moreover, the rare surface occurrences of even "late" Nebraskan tills (eg., the Iowa Point) makes correlations of till units older than 2.2 MYBP virtually impossible.

The problems of recognizing very old glacial deposits come into focus with a hypothetical example in which a "glacial advance" into east-central Illinois at around 4.0 MYBP deposited several meters of till before the ice melted. Suppose that a soil-forming interglacial interval followed this hypothetical advance. Because Nebraskan Till is represented very poorly in east-central Illinois, only the 4.0 MYBP deposits would underlie Kansan deposits, with presumably a well-developed partly eroded soil profile in between. The hypothetical soil would actually represent the post 4.0 MYBP interglacial, and the Aftonian but would likely be misinterpreted as Aftonian Soil on Nebraskan Till. This misinterpretation would be especially likely if only a few exposures were available for study as would probably be the case. The "old" till then would be correlated incorrectly with Nebraskan tills in the west-central part of the state.

Although the preceding paragraph is all supposition, a scenario of this sort must be considered when one is critically examining Nebraskan deposits, even in the case of Shimek's (1909) type Nebraskan.

In the supposedly unglaciated northwestern corner of Illinois, known as the "Driftless Area", there are actually scattered erratic boulders and other vague indications of glaciation that appear to be very old (certainly pre-Kansan) and that, without any other supportive evidence, have been tentatively considered to be Nebraskan in age (Frye et al. 1965). The old erratics of the "Driftless Area" could just as well represent the erosional remnants of a pre–Deadman Pass midcontinental glacial event. Unfortunately, there does not appear to be any way to adequately test this hypothesis nor to unambiguously extend the midcontinental glacial stratigraphic sequence to beyond about 3.0 MYBP.

Glacially Induced Sea-Level Fluctuations: Records from the Gulf Coastal Plain

Introduction

The expansion of glaciers on the continents will lower sea level about 10 meters for every 3.6×10^6 km^3 of accumulated ice. It has been estimated independently from oxygen isotope records that the growth of late Wisconsinan glaciers lowered sea level by perhaps as much as 160 meters (Shackleton 1977). In the preceding pages, North American midcontinent and mountain glacial stratigraphies extending back to about 3.0 MYBP have been reviewed. It is therefore appropriate at this point to examine the sea-level record for potential correlations to the terrestrial glacial record.

Late Cenozoic coastal plain sediments of Louisiana and Texas contain five shoreline terraces and related depositional surfaces that may correspond to interglacially high sea levels (Fisk 1944). In Louisiana, the terraces and related surfaces have been assigned from oldest (highest) to youngest (lowest) the names Williana, Bentley, Montgomery, Prairie, and Recent (Fisk 1939; 1944). Each of these lithostratigraphic surfaces is associated with an erosional-depositional sequence that began with valley cutting during an initial regression. The valley-cutting phase was then followed during an ensuing transgression by the accumulation of gravelly sand, grading upward into finer sands, silts, and clays (Bernard 1950). Fisk (1944) empirically correlated the valley-cutting episodes of the Williana, Bentley, Montgomery, Prairie, and Recent terraces, respectively, with the Nebraskan, Kansan, Illinoian, Early Wisconsinan, and Late Wisconsinan glacials, and the intervening depositional phases with the Aftonian, Yarmouthian, Sangamonian, and Peorian interglacial intervals. The Peorian, which originally separated the Iowan from the Wisconsinan and then was reduced in rank to a substage of the Wisconsinan is, however, no longer considered a valid stratigraphic term (Willman and Frye 1970).

Toward a Gulf Coast Terrace Chronology

The highest coastal plain terrace, the Williana, lies at elevations of about 200 to almost 600 ft. (60–180 m); the Bentley, between 100 and 200 ft. (30–60 m); the Montgomery, between 70 and 125 ft. (20–40 m); and the Prairie from a few inches above sea level to 100 ft. (30 m) (Bernard and LeBlanc 1965). The maximum uplift of each terrace should be proportional to its age, and if uplift is assumed constant, the ratio of 600 (Williana) to 200 (Bentley) to 125 (Montgomery) to 100 (Prairie) should approximately reflect the age differences of the valley-cutting episodes associated with the terraces. Thus the earliest Williana regression

(Nebraskan of Fisk 1944) is approximately six times older than the Prairie regression, whereas the Bentley regression (Kansan of Fisk, 1944) is only twice as old as the Prairie.

Akers and Holk (1957) in their comprehensive study of the subsurface stratigraphy of the Gulf Coast traced the Williana, Bentley, Montgomery and Prairie formations through a series of onshore and offshore drilled wells in southern Louisiana. Their data show the average thicknesses of the formations to be about 800 ft. (245 m) for the Williana; about 600 ft. (180 m) for the Bentley; and about 200 ft. (60 m) each for the Montgomery and Prairie. These thicknesses correspond to a base depth of about 1800 ft. (550 m) for the Williana; 1000 ft. (300 m) for the Bentley; 400 ft. (120 m) for the Montgomery, and 200 ft. (60 m) for the Prairie. If average accumulation rates have been uniform through time the ratio of the maximum ages of each formation should be about 1800:1000:400:200, which reduces to 600:333:133:67. The similarity of the set of ratios reflecting terrace uplift (600:200:125:100) to the set of ratios reflecting subsidence and deposition (600:333:133:67) suggests that they may be averaged together and the average used quantitatively to estimate the relative ages of the formations. If any of the formations can be dated independently, the average ratio (reduced to 14:6:3:2) can be used to estimate the actual ages of the others.

It may be possible to date the Williana Formation from its biostratigraphic position in Gulf Coast wells which can be tied to the paleomagnetic time scale (see chaps. 1 and 2). According to Akers and Holk (1957), the Williana Formation is equivalent to the lower 1500 ft. (450 m) of an offshore Louisiana well that contains several planktonic foraminiferal and nannofossil datums (Poag 1972). Correlation of those microfossil datums to the paleomagnetic time scale has been summarized by Berggren (1973) and Berggren and Van Couvering (1974). Akers and Holk's correlation implies that deposition of the Williana Formation began prior to the extinction of *Globoquadrina altispira* about 2.8 MYBP and terminated some time after the first appearance of abundant *Globorotalia truncatulinoides*, about 1.8 MYBP.

If an approximate 2.8 MYBP age is taken as a minimum estimate for the initial Williana regression, the ratio 14:16:3:2 obtained above provides minimum age estimates of about 1.2 MYBP for the initial Bentley regression, 0.6 MYBP for the initial

Montgomery regression and 0.4 MYBP for the initial Prairie regression. These age estimates for the initiation of each of the Gulf Coastal Plain terrace formations are compared to the midcontinent glacial stratigraphic record in figure 5.6 (column 7, *arrows*).

A volcanic ash zone situated in the Williana formation has been traced in the subsurface by Murray (1961). It appears in well No. 10 of Akers and Holk (1957) at around 1750 ft. (530 m), i.e., in the lower half of the Williana Formation. This ash could be from either a western United States source, for example, a Pearlette-type ash from the Yellowstone volcanic province, or from a Central American source. It should be possible either to date the Williana ash or at least to correlate it with tephra layers of similar stratigraphic position by using trace elements (see chap. 4). For example, the late Pleistocene Los Chocoyos ash from Guatemala constitutes a disseminated, but nevertheless readily correlatable, layer in Gulf of Mexico marine sediments (Drexler et al. 1980 and unpublished data; see also chap. 4, this volume). The 2.8 MYBP estimated age of the initial Williana regression suggests that the Williana ash zone could be equivalent to the 2.2 MYBP ash zone in Iowa (Boellstorff 1978) and to the possibly identical type "B" Pearlette ash dated about 2.0 MYBP (Naeser et al. 1973).

Erosion related to the Williana regression about 2.8 MYBP is compatible with a major continental glacial event occurring at the same time as the Deadman Pass mountain glaciation in the Sierra Nevada. The 1.2 MYBP minimum age estimate for the initial Bentley regression is, in turn, compatible with the initiation of major continental glaciation just after deposition of the 1.2 MYBP type "S" Pearlette ash. The type "S" Pearlette ash was deposited either during the Aftonian interglacial (Berggren and Van Couvering 1974), or, more probably, during a mid- to late-Kansan interstadial (fig. 5.6, column 7). The 0.6 MYBP minimum age estimate for the initial Montgomery regression corresponds to the Yarmouthian interglacial, which contains the type "O" Pearlette ash. There is general agreement that the type "O" Pearlette ash is post-Kansan and pre-Illinoian (Reed et al. 1965; Cooke, 1972; Berggren and Van Couvering 1974). The initial Montgomery regression therefore was probably related to the Yarmouthian–early Illinoian (Liman) transition (fig. 5.6, column 7; fig. 5.4).

The initial Prairie regression is estimated to have occurred around 0.4 MYBP. This may indicate a major continental ice-volume buildup, one coincident with the Casa Diablo Mountain glaciation in the Sierra Nevada but distinct from the earliest Illinoian (e.g., the Monican substage—see fig. 5.4). A return to ice-free conditions after the initial Prairie regression is in accord with a Mindel-Riss interglacial occurring from about 0.4 to 0.3 MYBP, as was suggested by Berggren and Van Couvering (1974) (fig. 5.7). This "interglacial" would correlate either with an expanded Sangamon (e.g., Cooke 1972) or with a late Illinoian interstadial soil-forming episode (fig. 5.4).

European Glacial Stratigraphy: The Alps

Introduction

Penck was the first to recognize a series of fluvial gravel terraces and associated moraines in the Alps as indicators of major glacial events (Penck and Bruckner 1909). On this basis, he defined four glacial stages (Günz, Mindel, Riss, and Würm) related to the four lowest (youngest) and best expressed terraces. Intervening interglacials were not named separately but were designated as Günz-Mindel, Mindel-Riss and Riss-Würm (fig. 5.6, column 10). Penck's Alpine river-terrace glacial stages quickly gained wide acceptance because they could be easily correlated throughout drainage systems in the Alps (Eberl 1930). Much later, two higher and older terrace gravel deposits were added to the sequence: the Donau (Eberl 1930) and the Biber (Schaefer 1953) (fig. 5.6, column 10).

Each of the six major terrace stages recognized in the Alps began with significant downcutting of river valleys in the headwaters of the Danube (Donau) and its tributaries. Each episode of downcutting was followed by the deposition of thick gravel beds in the valleys (Richmond 1970; Kukla 1977). Renewed erosion at the initiation of the succeeding cycle then converted the gravel-floored valley surfaces into gravel-capped terraces overlooking the newly entrenched river valleys.

The Alpine terrace gravels were assumed for many years to have been deposited contemporaneously with periods of maximum glacier expansion in the Alps. Recent ^{14}C-dating of wood

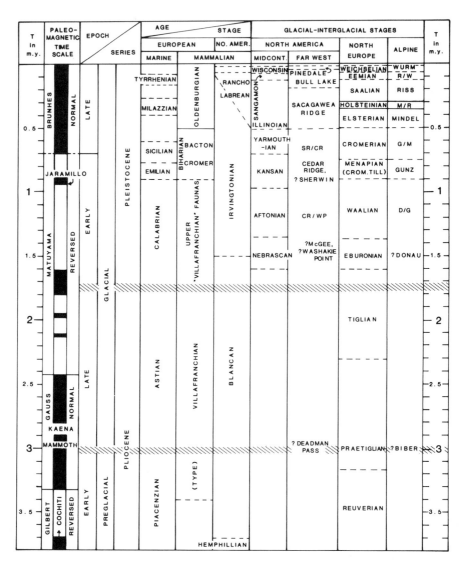

Figure 5.7 *A recently proposed chronology of late Pliocene/Pleistocene glacial and biostratigraphic events (condensed after Berggren and Van Couvering 1974).*

imbedded in the Würm gravels, however, has produced dates ranging from about 9,700 BP to about 400 BP (summarized in Kukla 1977). These dates clearly indicate that the Würm gravels were in part deposited during the late deglacial period and the recent interglacial. In a paleoglacial sense, therefore, the Alpine "glacial" stratigraphic stages that represent periods of gravel deposition are asymmetrically disposed with respect to Alpine glacial maxima. The Alpine "interglacial" stratigraphic stages (Riss-Würm, etc.), which represent downcutting episodes, may therefore be for the most part representative of an early glacial, rather than interglacial, environment (Schaefer 1953; reported in Kukla 1977).

Correlation of Alpine and North American Glacial Records

In Europe, tills and interglacial soils are known from the Günz, Mindel, Riss, and Würm stages but not from the Donau and Biber, which are represented solely by terraces (Richmond 1970). Richmond (1970) was able to draw correlations between glacial events in the Alps and the Rocky Mountains on the basis of similarities of till weathering, landscape erosion and soil development. The younger part is well documented by numerous ^{14}C dates. Richmond's correlation is further supported back to about 500 KYBP by a comparison of K-Ar dates on volcanic rocks from the Yellowstone region with dates from the Laacher volcanic province in the Rhine Valley (Richmond 1970) (fig. 5.6, column 10). Because the Rocky Mountain glacial stratigraphy can be correlated, in part, with the midcontinental North American glacial stratigraphic sequence, this effectively provides a trans-Atlantic linkage between the classical North American and European stratigraphic schemes back to Günz (late Kansan) time (Cooke 1972; Berggren and Van Couvering 1974).

Pre-Günz Glacials?

The European glacial stages that precede the Günz have not been dated radiometrically. Long-distance correlations therefore are problematic. What is known about the stratigraphic relationships of the older terrace deposits is based solely on their relative heights and associated faunas.

Donau till has not been clearly identified anywhere in Europe. Donau gravels are considered glacial, however, because they are bouldery, contain faceted stones (some of erratic provenance), and yield arctic molluscs (Schroder and Dehm 1951; reported in Richmond 1970). Biber gravels also contain arctic molluscs (Stromer and Lebling 1929; reported in Richmond 1970) and may also have been derived from now missing or still undiscovered glacial deposits. Richmond (1970) and Berggren and Van Couvering (1974) both correlate Donau with Washakie Point–Nebraskan. Richmond based his correlations on similar weathering characteristics and heights above the valley floors, while Berggren and Van Couvering based their correlations on mammalian faunal zones (fig. 5.7). Berggren and Van Couvering (1974) then go on tentatively to correlate the Biber with an early Blancan (North American mammal stage) cold-climate phase and in turn with the Deadman Pass glaciation of the Sierra Nevada (fig. 5.6).

According to Richmond (1970, p. 7) the Biber gravels "lie on a series of as many as six high erosional terraces". The evidence available to date unfortunately cannot resolve the question as to whether the Biber spans a period equal to or greater than all of post-Günz time (500,000–1.0 million years), in which only four major terraces developed. An affirmative answer could make the oldest Biber roughly about the same age as the Deadman Pass Till or about 3.0 million years old (fig. 5.6, column 10). The key question is whether each of the six Biber terraces represents an individual Alpine glacial cycle of stage rank. This represents a problem like trying to define the base of the Nebraskan in North America. It may never be satisfactorily solved by field observations within the formerly glaciated regions, but it certainly indicates the possibility of diverse and extensive pre-Kansan–Günz temperate mountain glaciations, perhaps even predating 3.0 MYBP.

In a recent attempt to correlate the classical Alpine terraces with warm-cool fluctuations in floras and molluscan faunas recorded in paleomagnetically dated sediments of the lower Rhine Basin, Brunnacker et al. (1982) placed the oldest Günz, at least 300,000 years younger than earlier workers had done. Furthermore, their earliest evidence of a cold glacial climate in the lower Rhine Basin is in Praetiglian-equivalent sediments containing what is probably the Reunion magnetozone, which dates from about 2.0 MYBP (Mankinen and Dalrymple 1979). Brunnacker et al. (1982) consider that this early cold period marks the advent of Alpine glaciation, and they therefore assign the oldest Biber gravels an age of about 2.0 MYBP. Their alternative stratigraphy is indicated in brackets in figure 5.6 (column 10). If correct, it implies that early North American continental and mountain glaciations predated the first European glaciation by approximately 1.0 million years.

European Continental Glacial Stratigraphy

The Northern-European Plain

In northern Europe, ice advancing from the Scandinavian region crossed the North Sea into the British Isles, the Netherlands, northern Germany, Poland, and the Ukraine, leaving a series of at least three major till sheets and associated moraines. Through long usage, these glacial deposits became known (from oldest to youngest) as Elster, Saale, and Weichsel. Intervening peat layers containing temperate pollen, mollusc and mammalian fossil assemblages provided the basis for naming interglacials: Holsteinian (between Elster and Saale) and Eemian (between the Saale and Weichsel) (fig. 5.6, column 11). Eventually, a fourth series of glacial deposits, the Warthe, was recognized as distinct from Weichsel and Saale (Picard 1964) (fig. 5.6, column 11). The Warthe glaciation is older than Eemian but a "new" interglacial between the Saale and the Warthe has not yet been named. Kukla (1977) notes that some so-called Eemian interglacial sections are probably pre-Warthe and post-Saale in age.

All of the above glacial and interglacial stages have type areas on the north European mainland. The existence of a pre-Elster interglacial, the Cromerian, however, has been inferred from a type section in the British Isles that includes the Cromer Forest Bed of East Anglia (Turner 1975). The Cromer Forest Bed is overlain by the Lowestoff Till and underlain by the Cromer Till, the oldest till in East Anglia. The Lowestoff Till is thought to be equivalent to Elster (Kukla 1977) (fig. 5.6, column 11). The Cromer Till therefore presumably represents a glacial advance during a pre-Elster glacial stage. The only evidence for such an event on the continent of Europe is inconclusive and comes from borehole sections in Poland (Mojski and Ruhle 1965), where a possible pre-Elster glacial event has been named the Elbe glaciation (fig. 5.6, column 11).

In Germany, pre-Elster interglacial lake sediments have been identified that resemble the type Cromerian in their pollen assemblage. These deposits contain a paleomagnetic reversal that appears to record the Brunhes/Matuyama paleomagnetic boundary (H. Miller personal communication in Kukla 1977). The Brunhes/Matuyama boundary is dated at 0.73 MYBP. (Mankinen and Dalrymple 1979). These findings imply that the Cromerian interglacial is roughly equivalent to the Yarmouthian of North America and the Günz-Mindel of the Alps (fig. 5.6).

Non-glacial Paleoclimatic Records from Northern and Eastern Europe
Evidence from Pollen

In the Netherlands, nonglacial organic sediments older than the Cromerian interglacial have been divided into a series of generally "warm" or generally "cold" stages on the basis of pollen assemblages (Zagwijn 1960) (fig. 5.6, column 11).

The oldest pollen zone, the "cold" Menapian, probably includes the "Cromer Till glacial event" of East Anglia (Berggren and Van Couvering 1974) (= Elbe? in fig. 5.6). The cold-Menapian pollen zone probably correlates with the late Kansan or Günz glacial, the warm Waalian pollen zone with a middle Kansan interstade, and the cold Eburonian pollen zone with the early Kansan or Donau glacial. The cool-Praetiglian pollen zone may correspond in part to the Deadman Pass glacial of the Sierra Nevada, to the Biber of the Alps and to the early(?) Nebraskan of midcontinental North America (fig. 5.6). This interpretation is generally supported by intercorrelation with European and North American mammalian zones (Berggren and Van Couvering 1974) (fig. 5.7).

Paleoecological interpretations of pre-Menapian (= pre-Elbe) pollen sections from the Netherlands and France have added greatly to our understanding of the late Cenozoic paleoclimatic history of Europe (Zagwin 1975). The earliest cool phase following the end of the Miocene appears to correspond with the base of the Reuverian which "could be interpreted as boreal" (Zagwin 1975) (fig. 5.6, column 11). There are also clear indications of at least one other boreal event in the Reuverian. A boreal climate in the Netherlands shortly after the end of the Miocene (5.0 MYBP) could well mean a significant climatic deterioration compared to the modern interglacial climate, perhaps even including glaciation in the Alps and Scandinavia.

Zagwijn (1975) also indicates that the Praetiglian was the first time a tundralike vegetation characterized in the Netherlands. The Tiglian pollen zone (fig. 5.6, column 11), which succeeds the Praetiglian, appears generally warmer than the Praetiglian but contains at least one major tundra event. The Eburonian varies between tundra and boreal: the Waalian between temperate forest and boreal.

Wijmstra (in van der Hammen et al. 1971), continues the preglacial pollen record discussed by Zagwijn (1975) with a long glacial-period pollen profile from unglaciated eastern Macedonia which spans Cromerian to Recent. In Macedonia, the Elsterian, Saalian, and Weichselian glacial periods were generally characterized by tundralike conditions; the Cromerian, Holsteinian, and Eemian interglacial periods, by oak forest.

Evidence from Loess Deposits

During northern-European and Alpine glaciations, the prevailing westerly winds blew silt eastward, depositing it as loess in the principal river valleys of Czechoslovakia, Austria, and the Ukraine. With the retreat of the ice, forests grew and soils developed on the loess. Loess deposits and buried paleosols in eastern Europe therefore have long been recognized as valuable indicators of European glacial-interglacial events (Penck and Bruckner 1909). Loess sequences can be dated using paleomagnetic stratigraphy, thereby allowing sections from different localities to be patched together. On the basis of paleomagnetic polarity, invertebrate and vertebrate fossil assemblages, and lithostratigraphy (Kukla 1977), the resulting recording of loess- and soil-forming intervals can be correlated down the Elbe Valley to northern-European glacial and interglacial deposits, or up the Danube Valley to the Alpine terraces.

Kukla (1977) recognizes 22 soil-loess cycles (labeled V through A) in sections at Brno, Czechoslovakia, and at Krems, Austria. Figure 5.8 summarizes his correlation of loess cycles K through B to the classical Alpine terraces. The loess cycles K and V (the oldest) occur within or just below the Jaramillo and Olduvai subchrons respectively. Cycle K therefore, has an age of approximately 0.9 MYBP, and cycle V, an age of about 1.7 MYBP

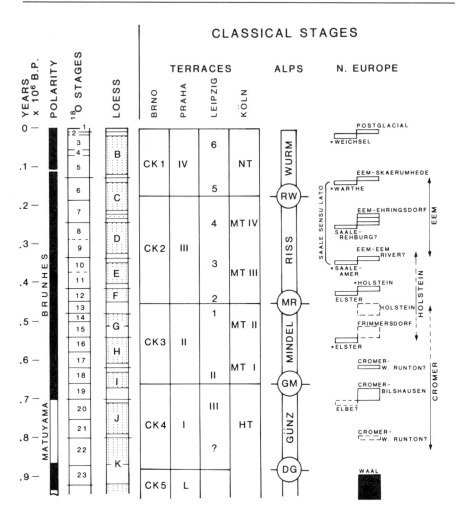

Figure 5.8 *A suggested correlation of oxygen isotope stages, loess cycles, and European glacial stages (con-densed from Kukla 1977).*

(Mankinen and Dalrymple 1979). This yields an average duration of about 82,000 years for each complete cycle (cycle A is now represented only by a soil horizon) and 100,000 years for the 17 cycles considered by Kukla to include glaciations as severe as the Weichsel. The latter figure is very close to the average period of isotopically defined glacial events in the deep-sea record. This similarity led Kukla (1977) to date individual loess cycles by directly correlating the loess record to the chronologically well–controlled deep-sea record of $\delta^{18}O$ stages (fig. 5.8).

The Marine Record of Glacial Events

Introduction
Continental glaciations are a reflection of major perturbations in the Earth's hydrologic cycle. Water that evaporates from the sea surface is transported as atmospheric water vapor over the continents, where it condenses and falls as rain and snow. When, over periods of tens of hundreds of years, more snow accumulates than melts, glaciers and ice sheets are formed that slowly flow downhill, toward the sea. The original water molecules may thus return to the sea either in the form of icebergs calved from marine glacier termini or in flows of glacial meltwater.

The oceans exert an influence on continental glaciations because oceanic temperatures, currents, and wind patterns strongly affect the evaporative transport of moisture onto the continents. Conversely, glaciers and continental ice sheets exert a marked influence on the oceans.

The growth of continental glaciers causes a corresponding decrease in the volume and, therefore, in the level of the oceans. Recent estimates of the potential glacial sea-level lowering have ranged from about 160 m (Hughes et al. 1977) to as little as 60 m (Field et al. 1979; Blackwelder et al. 1979). Even a sea-level lowering of 60 m, however, would have a profound effect on continental-margin morphology and sedimentation. Icebergs, disintegrating in the sea, distribute coarse and fine glacial debris over vast areas (Ruddiman 1977). The icebergs also cause sea-surface cooling (Ruddiman and McIntyre 1981). During deglaciation, large injections of turbid glacial meltwater alter the temperature-salinity stratification and sediment budget of the oceans.

Given all the potential interactions between continental glaciers and the oceans, it is not surprising that shortly after the development of the first techniques that provided the capability of dating marine sediments (^{14}C and U-Th), paleoceanographic interpretations of deep-sea cores were published with a view toward linking the marine record to the history of continental glaciation (Emiliani 1955; 1961; 1964; 1966; and Ericson and Wollin 1956a; 1956b; 1968; 1970).

Efforts To Relate Oceanographic Records To Continental Glacial Records

Work to determine the isotopic ratio of $^{18}O/^{16}O$ in the calcium carbonate tests of foraminifera was pioneered by Emiliani (1955). The application of this technique for resolving changing global ice volumes is reviewed in detail in chapter 2. Emiliani found a clear indication of variation between glacial and interglacial states, with a rather uniform periodicity of about 90,000 years. These alternating glacial and interglacial isotope stages today constitute the basis of a global oxygen isotope stratigraphy (see chap. 2).

At nearly the same time as Emiliani's initial isotope work was published, Ericson and Wollin (1956) designated alternating zones of presence and absence of *Globorotalia menardii* in Pleistocene deep-sea cores. The zones, lettered Q through Z (Ericson and Wollin 1968), showed that five complete "warm" to "cold" cycles were present within the last 1.7 million years (fig. 5.6, column 2). Ericson and Wollin correlated their zonation to the classical mid-continent terrestrial glacial-interglacial sequence. Problems with their correlation (fig. 5.9) are discussed in the final section of this chapter, but their zonation remains a most useful biostratigraphic tool (see chap. 1).

By using a more sensitive multiforaminiferal species approach, Ruddiman and McIntyre (1976) and Briskin and Berggren (1975) (fig. 5.6, column 4) produced detailed paleotemperature curves for most of the Brunhes Chron (600 KYBP to the present). They employed the Ericson zones as a stratigraphic aid. The paleotemperature record that Ruddiman and McIntyre produced for the North Atlantic shows a frequency of variation very similar to that exhibited by ^{18}O curves (fig. 5.10).

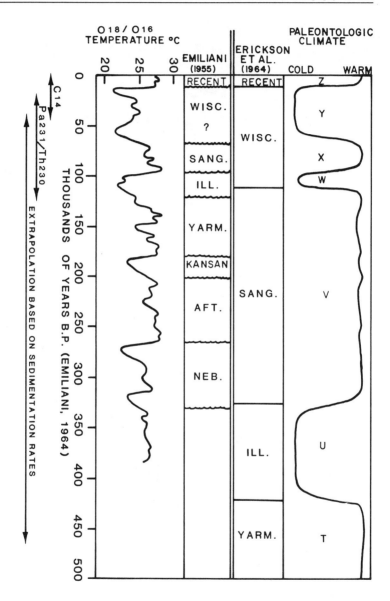

Figure 5.9 *A comparison of correlations of North American glacial/interglacial stages with an oxygen isotope record (Emiliani 1955, 1964) and with variations in the relative abundance of* Globorotalia menardii *(Ericson et al. 1964). (From Broecker 1965. Reprinted with permission of Princeton University Press.)*

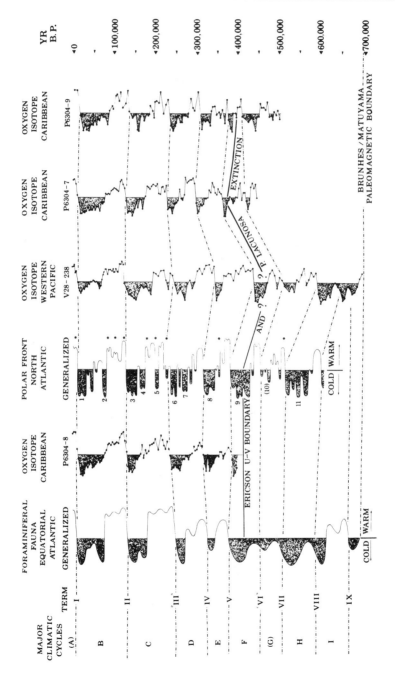

Figure 5.10 *Comparison of cyclic paleoclimatic records in deep-sea cores. (From Ruddiman and McIntyre 1976. Reprinted by permission of the authors.)*

Iceberg-rafted debris (IRD) in deep-sea cores has been studied in the Southern Ocean around Antarctica (Fillon 1977), in the North Atlantic (Ruddiman 1977; Fillon et al. 1981); and in the North Pacific (Kent et al. 1971). The presence of IRD (typically taken as inorganic detrital material coarser than 63 μm) clearly indicates the former presence of debris-laden icebergs; however, the amount of IRD will not necessarily give a straightforward indication of the amount or extent of glacier ice on the continents. The picture of IRD discharge is further complicated by the fact that IRD is common in some areas even during interglacials, as is indicated by the modern iceberg flux. Von Huene et al. (1976), working in the northeast Pacific beyond modern iceberg limits, delineated an 800,000-year-long record of sand and pebble abundance that strongly mimics deep-sea $\delta^{18}O$ and paleotemperature curves from the same period (fig. 5.11, compare with fig. 5.10). This work provides direct evidence that at least coastal Alaskan mountain glaciers may have fluctuated with the same periodicity as sea-surface temperatures and the oxygen isotope record.

Marine Evidence of Pre–2.0-MYBP Climate

Berggren (1972) defined the beginning of iceberg rafting in the North Atlantic on the basis of studies of three DSDP sites (111–Orphan Knoll at 50°26′N, 46°22′W; 112–south central Labrador Sea at 54°1′N, 46°36′W; 116–Rockall Bank at 57°30′N, 15°56′W). He recorded a change from virtually zero detrital grains to abundant detrital grains (clays excluded) at a core depth corresponding to approximately 3.0 MYBP. This level, which actually could be as young as 2.5 MYBP (Backman 1979), also corresponds to a change from foraminiferal and coccolith carbonate sediment to a dominantly terrigenous sediment and to a change from a Gulf Stream fauna to a subpolar fauna.

The present (interglacial) course of the Gulf Stream, up the eastern coast of the United States, and northeastward across the North Atlantic to Great Britain, is virtually free from melting icebergs. Therefore Berggren's (1972) data permits one to conclude that a Gulf Stream–like flow of warm surface water was a persistent feature of the southern Labrador Sea and northeastern North Atlantic prior to 3.0 MYBP and that if cold, iceberg-

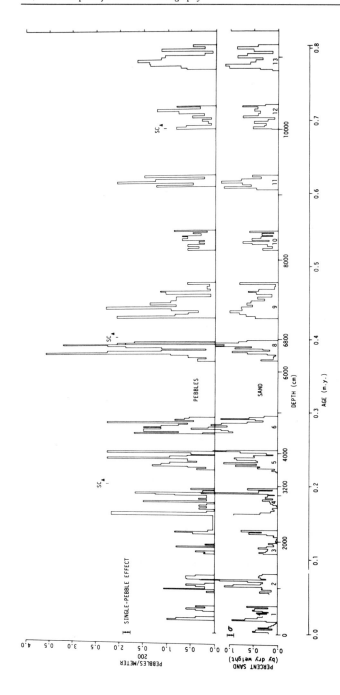

Figure 5.11 Percent sand and pebbles per meter are both measures of ice-rafted debris content in sediments cored at DSDP Site 178 in the North Pacific. (From von Huene et al. 1976. Reprinted by permission of the authors.)

bearing currents existed they did not pass over DSDP Sites 111, 112, and 116.

Recently, Warnke (1982) reported evidence of ice rafting going back to about 5.0 MYBP in the Norwegian and Greenland seas. Therefore, significant temperate continental glaciation of parts of Europe, Asia and North America in the period 5.0 MYBP to 3.0 MYBP cannot be ruled out on the basis of IRD distribution alone.

Oxygen isotope records spanning the interval from about 4.0 MYBP to about 2.0 MYBP (Shackleton and Opdyke 1977; Howell 1982) suggest that there were substantial increases in mean global ice volume at about 3.2 MYBP, and again at about 2.5 MYBP. Shackleton and Opdyke (1977) inferred that these base-level changes were probably due to variations in Northern Hemisphere ice volume. Deep-sea isotope records, however, cannot distinguish between ice stored in continental ice sheets and ice stored in floating, marine-based ice shelves, which may have attained an appreciable thickness in the Arctic Ocean (Broecker 1975; Williams et al. 1981; and Fillon and Williams 1983). In regard to $\delta^{18}O$ variations in the period prior to 3.2 MYBP, Shackleton and Opdyke (1977, p. 218) state that "it is conceivable that the variability stems entirely from analytical error" but they also acknowledge that "if they [the variations] derive from real glacial events, they could be caused by changes in the Antarctic Ice Sheet or by small Northern Hemisphere glaciations".

A conservative assessment of the deep-sea marine evidence so far presented leads to three important conclusions: (1) small (in volume but not necessarily in area) Northern Hemisphere, temperate-latitude continental glaciers existing prior to 3.2 MYBP cannot be ruled out; (2) Northern Hemisphere continental ice sheets probably did not reach their maximum volume and extent until sometime after 3.2 MYBP, and probably not until after 2.5 MYBP; and (3) the longevity of individual Northern Hemisphere continental ice sheets in temperate regions has probably been on the order of 80,000 to 100,000 years or less.

The Gulf of Mexico

Observations of foraminiferal faunas and lithologic changes in late Cenozoic Gulf of Mexico sediment cores from continental shelf and slope locations provide additional evidence of possibly glacier-related paleoenvironmental variation. The resolution, continuity, and chronology of much of this work, however, requires some refinement. A case in point is a recent study of slope core E67–135 from the DeSoto Canyon (Brunner and Keigwin 1981). Brunner was able to recognize five zones of *Globorotalia menardii* abundance, and so indicated the presence of Ericson zones Q through Z (fig. 5.6, column 2). Their zonation was refined subsequently by Neff (1983) (see chap. 2, this volume). The isotope study of E67–135 done by Keigwin resolved only eight interglacial stages (low $\delta^{18}O$ values) above zone Q, where there should have been 20 (Van Donk 1976). What happened to the 12 missing interglacials? Two possible explanations for their absence are (1) an insufficient number of isotopic analyses failed to resolve several interglacial events; and (2) the *G. menardii* zones in E67–135 represent incomplete Ericson zones in which a number of isotope stages have been lost owing to small unconformities or to intrazonal intervals of very low sedimentation rate. The biostratigraphy of this core is discussed in greater detail in chapter 2.

Another example of apparent unconformities in Gulf of Mexico continental-margin sediments deposited during the last 2.0 million years comes from a Gulf of Mexico lower-continental-slope core, E67–125 (Beard et al. 1982). Data presented for that core include paleomagnetic polarity, *G. menardii* abundance, and the extinction level of the coccolith *Pseudoemiliania lacunosa* (fig. 5.12). Gulf of Mexico bathymetric cycles inferred chiefly from seismic profiles (Sangree 1978) are compared with a planktonic foraminiferal paleotemperature curve derived for the core, and with continental glacial stratigraphy (Beard et al. 1982) (fig. 5.13).

Van Donk (1976) and others place the *P. lacunosa* extinction at approximately the top of Ericson's U zone and the Olduvai magnetozone coincident with Ericson's P zone. In core E67–125, if Ericson's zones are assigned to *G. menardii* maxima using the top of the core, the *P. lacunosa* extinction and the Olduvai magnetozone as guides we find that the *G. menardii* peak above the *P. lacunosa* extinction is probably zone X, and that the next "warm" zone, V, must be missing (fig. 5.12). The abundance maxima, coincident with and below the *P. lacunosa* extinction, fall within the early part of the Brunhes magnetozone (perhaps partly within

Figure 5.12 *Geomagnetic polarity,* Globorotalia menardii *abundance and the last appearance datum (LAD) of* Pseudoemiliania lacunosa *and* Globorotalia miocenica *in northern Gulf of Mexico slope core E67–125 (after Beard et al. 1982). Suggested G. menardii zone and magnetozone identifications are added (this chapter).*

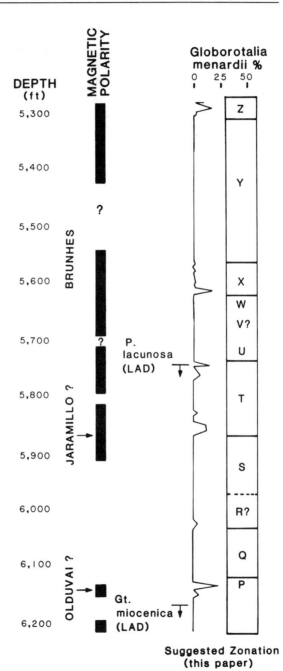

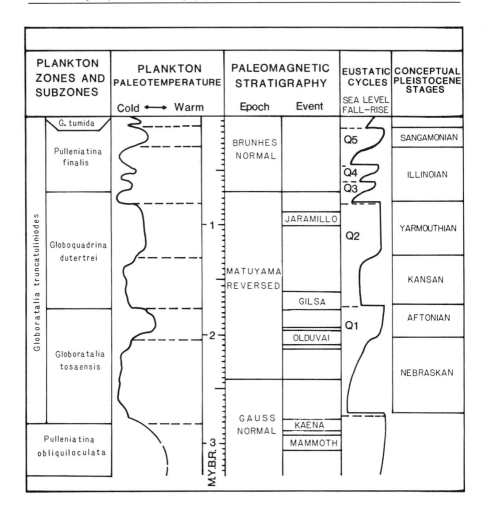

Figure 5.13 *Pleistocene eustatic cycles compared with paleomagnetic, planktonic foraminiferal and glacial stratigraphies (condensed from Beard et al. 1982).*

the Jaramillo magnetozone). They are thus in a position to be included in Ericson's T zone (Van Donk 1976). The Olduvai magnetozone, reached at the bottom of the section, includes the next-lower *G. menardii* maximum, the extinction of *Globorotalia miocenica*, and marks a coiling change in *G. menardii*. It therefore must coincide approximately with zone P. The warm R zone, if present, is not very well expressed (fig. 5.12).

The incompleteness of the Ericson zonation in core E67–125 (fig. 5.12) suggests that the foraminiferal paleotemperature curve presented by Beard et al. (1982) is based on a section that is relatively incomplete. Whether this is due to unconformities or to very low sedimentation rates during certain episodes cannot be resolved without recourse to a detailed isotope record for the core. Nevertheless, the match among the paleotemperature curve, the paleobathymetric records, and the continental glacial record (fig. 5.13) as demonstrated by Beard et al. (1982) offers the possibility that unconformities and/or major changes in sedimentation rate along the continental margin of the northern Gulf of Mexico may in themselves record a response to the principal continental glacial-interglacial alternations.

Up to the present, published studies of Gulf of Mexico cores have been based on rather widely spaced samples, as compared to other deep-sea-core studies. It is perhaps not surprising then that Gulf of Mexico core records of climatic and glacial changes appear to more closely resemble the incomplete continental glacial record than do the relatively continuous deep-sea records (fig. 5.6).

An Integrated View of Continental and Marine Events

Evaluating Continental-Marine Correlations
Many authors have attempted to draw correlations between the glacial events recorded in continental and marine records (Emiliani 1955; Ericson et al. 1964; Shackleton 1969; Cooke 1972; Berggren and Van Couvering 1974; Ruddiman and McIntyre 1976; van Donk 1976; Kukla 1977; Mangerud et al. 1979; Beard et al. 1982; Woillard and Mook 1982; among others). The best

correlation schemes are based on some combination of the following: terrestrial and marine faunal zones, K-Ar dating of lavas and volcanic ashes, radiocarbon dating, paleomagnetic stratigraphy, oxygen isotope stratigraphy, and uranium disequilibrium series dating on raised coral terraces. Figure 5.6 thus represents a compilation of the best available information from several continental and marine correlation schemes for the last 3.5 MYBP.

Figure 5.6 shows a strong consistency in placement of glacials and interglacials with respect to the time scale. All of the continental stratigraphies have in common interglacial events centered in the intervals from around: 100 KYBP to 300 KYBP, 600 KYBP to 800 KYBP, 1.0 MYBP to 1.4 MYBP, and 1.9 MYBP to 2.4 MYBP. Because those interglacial periods are tied to a reliable chronology, they can be compared directly with marine events. There does not appear to be any obvious correlation of interglacial periods with the isotope record of ice volume (fig. 5.6, columns 7–12), yet there is fairly good agreement with the Beard et al. (1982) eustatic cycles (fig. 5.6, column 4). Briskin and Berggren's (1975) sea-surface temperature fluctuations (fig. 5.6, column 4) and Ericson's and Wollin's (1968) "warm" *G. menardii* abundance maxima (fig. 5.6, column 2) cannot be said to correlate consistently with interglacials, and, in fact, they correlate more readily with glacials prior to zone W.

Of course, there are uncertainties inherent in placing boundaries in figure 5.6 and in assigning stage names to dated sections distant from type localities. Boundary uncertainties represented by diagonal stage boundaries in figure 5.6 are primarily a function of the paucity of chronologic control in certain geographic areas and parts of the stratigraphic section. Stage-name uncertainties represented by alternate stage names in figure 5.6 (set in brackets) are a function of the difficulty of physically correlating a datable glacial or interglacial section in one locality with a specific type section. Thus, Cooke (1972) referred to interglacial deposits reliably estimated to show an age of about 250 KYBP as Holsteinian, whereas Kukla (1977) pointed out that some interglacial deposits traditionally included in the Eemian appear to have an age of at least 250 KYBP. In such cases, the interglacial dates are in agreement, but the stage names are not.

Some further confusion stems from the way the Alpine terrace stages are interpreted climatically and whether the original definition of the nomenclature is used (Kukla 1977) or a general-usage nomenclature is employed (Cooke 1972; Berggren and Van Couvering 1974). The intervals between terrace-gravel depositional episodes (Riss-Würm, etc.) in Kukla's (1977) scheme are glacial. The terrace stage names, however, generally have been used to refer to glacials that are not necessarily fully contemporaneous with the type terrace gravels.

The large uncertainty in placing the Mindel-Riss "interglacial" (fig. 5.6) is due in part to the complexity of correlating dateable basalt pebbles in Rhine Valley gravels to valley incision on the Danube, and from there to the well-developed soil in the Alps, which is thought to represent a Mindel-Riss period of interglacial climate. This correlation (see discussion by Richmond 1970) produces the younger age limit shown in figure 5.6. The older age limit stems from a correlation of deep-sea $\delta^{18}O$ cycles to loess-soil cycles, which, in turn, are correlated to Danube terraces (Kukla 1977).

Search for the Common Denominator: Confirming the Continental-Marine Link

The problem to be grappled with in relating continental events to the deep sea is not simply one of determining how best to juxtapose the classical glacial record with the marine record in the proper time frame; it is of understanding the relationship between the two records. The place to seek enlightenment on this subject is in the nonglacial continental record of datable pollen and loess sequences. These records show a periodicity very similar to $\delta^{18}O$ and foraminiferal paleotemperature records from deep-sea sediments. As with marine records, pollen and loess records are not clearly related to the classical glacial-interglacial sequences. Because they are terrestrial, however, a direct linkage between them and marine records would imply a definite mechanism of interaction between the marine and terrestrial environments as is suggested by the hydrologic cycle.

Woillard (in Woillard and Mook 1982); Wijmstra (in van der Hammen 1971); and Kukla (1977) have provided terrestrial records of sufficient quality and completeness to establish a firm land-sea correlation. Similarities between Woillard's ^{14}C-dated pollen profile from northeastern France (fig. 5.14) and Ruddiman and McIntyre's (1981) $\delta^{18}O$ record from the central subpolar

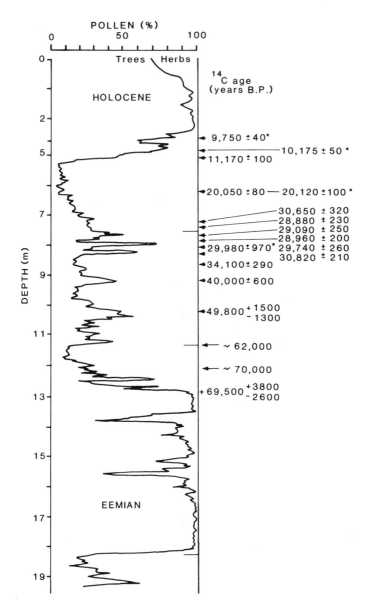

Figure 5.14 *Dated record of pollen assemblage variations in a core from the Grand Pile peat bog in Belgium (after Woillard and Mook 1982). Temperate conditions are reflected by a dominance of tree pollen.*

North Atlantic (core V30–97) (fig. 5.15) are particularly striking for the period 140 KYBP to 0 BP where tree-pollen maxima correspond to isotopic ice-volume minima. At a longer scale, Wijmstra's pollen profile from Macedonia (van der Hammen 1971) exhibits a very close similarity to Ruddiman and McIntyre's (1976) record of foraminiferal paleotemperature in central North Atlantic core K708–7 (fig. 5.16). In this case, oak maxima correspond to warm ocean temperatures. Literal correlation between the initiation of each oak maximum and the $\delta^{18}O$ defined terminations of glacials in the marine record places the base of the pollen record at the isotope stage 18/17 boundary, which is dated at approximately 647 KYBP (Shackleton and Opdyke 1976). This date agrees nicely with the Upper Cromerian age cited in van der Hammen et al. (1971).

Kukla's (1977) loess-soil cycles mimic the deep-sea $\delta^{18}O$ record for the last 140,000 years (fig. 5.17), and a careful compilation of loess-soil cycles during the last 1.0 million years (dated paleomagnetically) produces precisely the same number of cycles as are in the marine isotope record (Kukla 1977) (fig. 5.8). It seems fairly safe to conclude then that a mechanical linkage must exist between the marine and terrestrial environments.

Orbital Variations and Climate Change

Since the introduction of Milankovitch's (1941) orbital-mechanics theory of climate change, scientists have speculated on how variations in incoming solar radiation (insolation) caused by cyclical variations in the Earth's orbital parameters might be related to glacial-interglacial climate change. For a review of the insolation argument, see Zeuner (1945); Berger (1976, 1978); Hays et al. (1976); Kominz et al. (1979); Briskin and Harrell (1980); and Pisias and Moore (1981). These studies provide a wealth of evidence in favor of a definite linkage between deep-sea isotope and paleotemperature records and the orbital parameters controlling insolation, although the mechanisms that might be involved are not clarified.

The first plausible mechanistic theory of continental glacier control by orbitally induced insolation variations was presented by Ruddiman and McIntyre (1981). In essence, their model is able to explain glacial fluctuations at about the 23,000-year orbital periodicity (Berger 1978) in terms of feedback effects be-

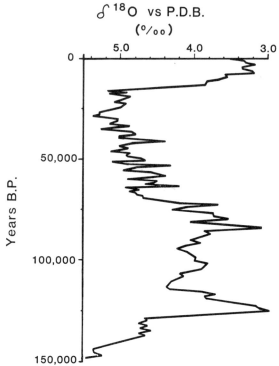

$\delta^{18}O$ vs P.D.B.
($^o/_{oo}$)

Years B.P.

Figure 5.15 *Oxygen isotope record from central subpolar North Atlantic piston core V30–97 plotted versus age. Greater ice volume to the left (after Ruddiman and McIntyre 1981). Note the similarity with the European pollen record in figure 5.14.*

Figure 5.16 *Foraminiferal assemblages and isotopic stratigraphy from North Atlantic core K708–7 compared with a pollen record and pollen zonation from the Tanaghi-Phillipon peat bog in Macedenia (after Kukla 1977). The base of the record is at about 647,000 BP (Shackleton and Opdyke 1976).*

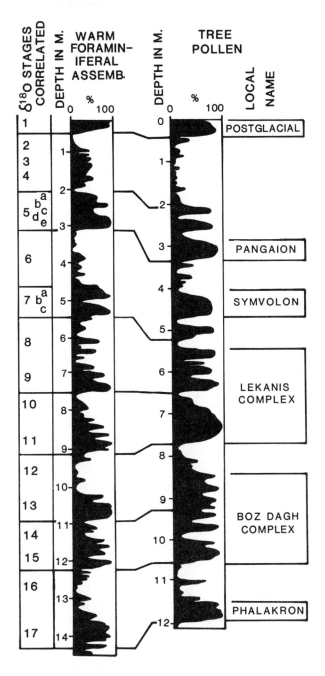

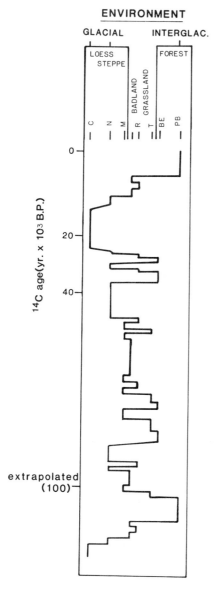

Figure 5.17 *Curve of reconstructed environment for approximately the last 130,000 years based on soil development in Czechoslovakian loess sections. C through PB are gradations of climate between cold loess steppe and warm forest (after Kukla 1977).*

tween sea-surface temperature and glacier growth or decay on the continents. If the 23,000-year orbital component of insolation variation can be directly linked to changes in continental ice volume and sea surface temperature, then it is probable that longer-period variations in the distribution of solar insolation on the Earth's surface also are causally linked through sea surface temperature to longer episodes of glacier expansion and retreat. The longer orbital periodicities which prevailed during the last 5.0 million years are 41,000 years, 96,000 years, and 413,000 years (Berger 1978; Briskin and Harrell 1980). Of these, the 41,000-year and 96,000-year periodicities are strongly evident in $\delta^{18}O$ and paleotemperature deep-sea records (Hays et al. 1976) and by analogy, therefore, in continental pollen and loess records. The 413,000-year periodicity is seen at best only weakly in deep-sea data (Briskin and Harrell 1980), possibly because continuous records spanning at least 2.0 million years are required to resolve periods of that duration.

Ericson Biostratigraphic Zones and Orbital Periodicities

If one examines Ericson and Wollin's G. *menardii* zones, it is evident that warm zones R, T and V, which span the interval from about 1.7 MYBP to about 150 KYBP are much longer than warm zones X and Z. The average periodicity of the three G. *menardii* abundance cycles R/Q, S/T, and U/V is about 500,000 years; that of cycles W/X and Y/Z, only about 75,000 years. This suggests that the G. *menardii* zonation may reflect a combination of two different frequencies of paleoceanographic change. The W, X, Y, and Z zones may exhibit a shorter G. *menardii* periodicity, because the cores examined in that time interval show higher sedimentation rates and therefore yield a higher-resolution curve of G. *menardii* abundance. Older zones, on the other hand, may detect only longer periodicities of insolation variation, because deep-sea cores that span the last 2.0 million years have very low sedimentation rates and yield low-resolution curves of G. *menardii* abundance. Similarly, the generally higher resolution available in younger terrestrial sections is responsible for the contrasting degrees of subdivision of the Wisconsinan and older glaciations. It is, therefore, easy to see why early one-for-one correlations between Ericson's deep-sea zonation and continental glacial events seemed to fit so well (Ericson et al. 1964).

It is now clear that zones abundant in G. *menardii* do not correlate precisely with interglacial events. When both are placed in sequence, according to their best possible chronologies (fig. 5.6), "warm" intervals of high G. *menardii* abundance (R–V) appear to coincide, for the most part, with glacial events. Because we know that sea surface temperatures vary dramatically, at periodicities of less than 100,000 years (fig. 5.10), the correlation of G. *menardii* (a warm-loving species) to glacials probably means that both the short, cold intervals and short, warm intervals within the long "warm" Ericson zones were of larger than average amplitude. During "cold" Ericson zones, on the other hand, the included shorter warmings and coolings were probably much less extreme. For example, the warmings were not quite warm enough for G. *menardii* to proliferate, and the coolings not quite severe enough to foster extensive continental ice sheets.

Hindcasting the Glacial Cycle

In work just completed (Fillon and Williams 1983), a new long-period orbital mechanics model of glacier growth is presented that does not assume that midlatitude continental ice is the only significant contributor to Northern Hemisphere ice volume. Arctic Ocean ice sheets are considered to have contributed significantly to the total ice volume.

Midlatitude insolation varies predominantly in response to the 23,000-year, 96,000-year and 413,000-year orbital periodicities, whereas high-latitude insolation varies predominantly in response to the 41,000-year; 96,000-year; and 413,000-year orbital periodicities (Berger 1978). Midlatitude and high-latitude Northern Hemisphere ice sheets therefore, should not expand and contract in phase with one another, yet isotope curves will see only total ice volume (Fillon and Williams 1983). Studies of continental glacial fluctuations, of course, record only variations in midlatitude ice sheets. Ocean temperature, on the other hand, plays an important role in regulating interaction between high- and midlatitude ice masses, because melting continental ice will cool the North Atlantic, and a cool northern North Atlantic would be a prerequisite for Arctic Ocean ice-sheet growth (Fillon and Williams 1983).

The Earth's orbital parameters have been calculated back to 5.0 MYBP (Berger 1976). It thus has been possible to compute an index of the likelihood of orbital conditions favoring formation of large Arctic Ocean ice sheets (A) and an index of the likelihood of orbital conditions favoring formation of large continental ice sheets (C) (Fillon and Williams 1983). Variations in these indices are plotted in figures 5.18–5.21 and are compared with the Ericson zonation, sea surface temperature record, eustatic cycles, continental glacial stratigraphy, and the oxygen isotope record in figure 5.6.

Arctic Ocean ice sheets require a cool North Atlantic, therefore an index of the tendency for Arctic Ocean ice sheet growth (A) (fig. 5.18) should correspond with cool sea surface temperatures and G. menardii minima. An index of the tendency for continental ice growth (C) (fig. 5.19) on the other hand, should correlate best with the record of continental glacial events. The combined record of (A) and (C) (fig. 5.20) should correlate with the oxygen isotope record by reflecting the tendency for large total ice volumes (fig. 5.21).

The dual Northern Hemisphere ice sheet theory implicit in the model presented by Fillon and Williams (1983) provides a reasonable explanation of why records of sea surface temperature (including the Ericson zones) correlate poorly with continental glacial records and why both of those records fail to correlate with the marine isotope record. First, there is a fundamental offset in the short-period response of high- and midlatitude ice sheets, namely 41,000 years vs. 23,000 years. Second, the apparent period of variation of any record is governed in part by the resolution at which the record can be studied. If deep-sea and terrestrial records are examined at grossly different resolutions, they will never be manifestly correlatable. In testing correlations between predicted values of (A), (C), and (A + C) and paleoenvironmental records it was therefore necessary to adjust the resolution of the (A), (C) and (A + C) curves individually to match that of the particular paleoenvironmental parameters being compared. Thus, if the curve of (C) is divided into glacials and interglacials at a resolution that reflects only the most intense interglacials, the result is a feasible approximation of continental glacial stratigraphies (fig. 5.6, column 6). Similarly, if the curve of (A) is viewed at a somewhat higher resolution, one compatible with that obtained in the very low-sedimentation rate deep-sea cores available to Ericson and Wollin (1968) and Briskin and Berggren

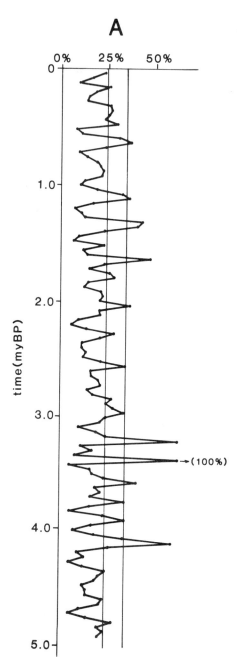

A

Figure 5.18 *The predicted tendency for cooler than normal arctic summers and a cold North Atlantic, conditions favorable for Arctic Ocean ice sheet growth (A); calculated from orbital parameters (after Fillon and Williams 1983). Lines at A = 24% and A = 34% indicate the limits used to define alternating "warm" and "cold" periods in figure 5.6, column 3.*

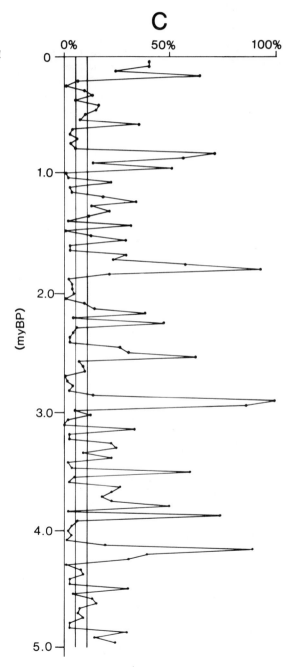

Figure 5.19 *The predicted tendency for cooler than normal midnorthern latitude summers favorable for continental ice sheet growth (C); calculated from orbital parameters (after Fillon and Williams 1983). Lines at C = 5% and C = 10% indicate the limits used to define alternating "glacial" and "interglacial" periods in figure 5.6, column 6.*

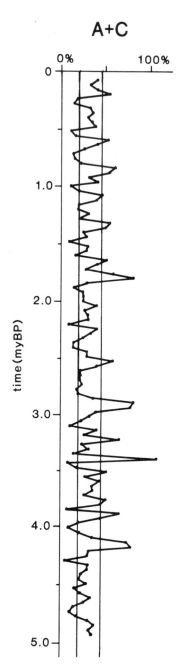

Figure 5.20 *The predicted tendency for coincidence of cooler than normal arctic and midnorthern latitude summers that would favor combined Arctic Ocean and continental ice sheet growth (A + C) (after Fillon and Williams 1983). Lines at A + C = 20% and A + C = 45% indicate the limits used to define alternating "glacial" and "interglacial" periods in figure 5.6, column 13.*

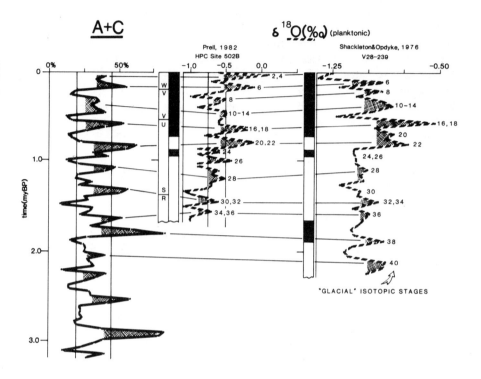

Figure 5.21 *Downcore, deep-sea oxygen isotope records block-averaged to yield representative isotopic values for ca. 40,000 year increments, compared with the curve of (A + C) (fig. 5.20) over the same period. The approximate positions of glacial oxygen isotope stages in the original (unaveraged) isotopic records are indicated (after Fillon and Williams 1983). Lines at $\delta^{18}O = -0.75$ ‰ and $\delta^{18}O = -0.5$ ‰ indicate limits used to define alternating isotopic "glacial" and "interglacial" periods in figure 5.6, column 12.*

(1975), it provides a feasible approximation of the Ericson zonation and the sea surface temperature record (fig. 5.6, column 3). Finally, high-resolution deep-sea oxygen isotope records should be best approximated by subdividing the curve of (A + C) into glacials and interglacials at a high resolution, one that reflects events having a wide range of predicted intensities, as in figure 5.6, column 13, and in figure 5.20. That appears to be the case for records of at least the last 1.0 MYBP (compare columns 12 and 13 in figure 5.6).

The dual Northern Hemisphere ice sheet theory can be further tested by examining a record of the magnitude of low-sea-level stands. Because sea-level will not be appreciably affected by the accumulation of a mostly floating ice-sheet in the Arctic Ocean (or in the circum-Antarctic Ocean), low-sea-level stands should record only variations in continental ice volume. Beard et al. (1982) provide a suitable sea-level record to compare with (C). Its chronology is based on paleomagnetic data and biostratigraphy (fig. 5.13; fig. 5.6, column 5). If Beard's "conceptual" glacial-interglacial stage names are removed from the sea-level record in figure 5.13, a general correspondence of the sea-level record with the continental glacial record presented in figure 5.6 is evident. There is also a resemblance between the sea-level and (C) records (see fig. 5.22).

A New Direction for Continental-Margin Studies
Recent advances in oxygen isotope stratigraphy and planktonic foraminiferal paleoecology/biostratigraphy in Pleistocene marine sediments are readily applicable to many continental-margin sediments. These techniques provide an effective means of tying continental-margin sections to deep-sea cores that have been dated, in absolute terms, by reference to paleomagnetic reversal stratigraphy. Up to the present, work on continental-margin sections has resolved microfaunal and sedimentologic variation that appears to respond in phase with the longer-period classical glacial-interglacial alternations of the continents rather than with the short climatic periodicities that dominate most deep-sea records. The long-period climatic variability thus observed, when supplemented with high-resolution oxygen isotope and microfaunal studies, may provide a key with which to reconcile the evidence of only three major continental glacial events with the

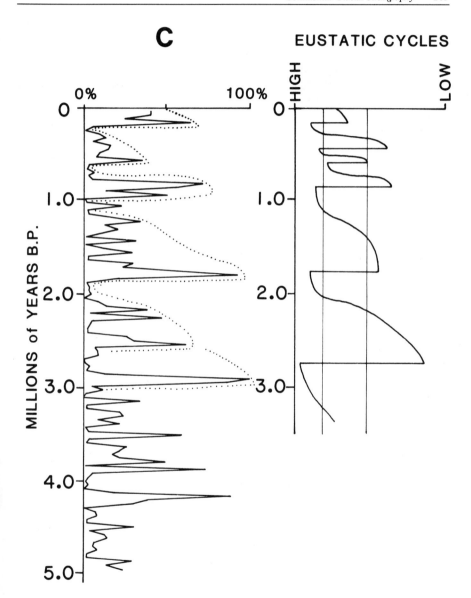

Figure 5.22 *Curve of C (fig. 5.19) compared with published eustatic cycles (sea level record). Dotted lines suggest long-term trends in orbitally predicted continental ice volume (adapted from Beard et al. 1982).*

evidence of 12 major global ice-volume maxima during the last 1.0 million years. In the former case, approximately 2¼ glacial-interglacial cycles are represented (Kansan through Recent), yielding an average periodicity of about 450,000 years. In the latter case, the average periodicity is about 91,000 years, for 11 full ice-volume cycles. Both the continental and deep-sea records therefore appear to be varying in response to known orbital frequencies (about 413,000 and 96,000 years) that control insolation variations, although each evidently responds most strongly to different periodic elements of insolation. Realization that both continental and deep-sea paleoclimatic records may be "correct" should encourage Pliocene/Pleistocene stratigraphers to work toward refining continental margin stratigraphies and paleoclimatic-paleoceanographic histories. These same records should also encourage climate modelers to tackle the difficult problem of terrestrial-marine contrasts.

Summary

Magnetostratigraphy provides an ideal set of globally synchronous datum levels for a first-order stratigraphic framework in Pleistocene sediments. Within this well-recognized framework, repeated sedimentary sections and unconformities may be recognized, sedimentation rates may be determined, and seismic reflectors dated. Radiometric dating of terrestrial lava sequences firmly fixes the absolute ages of the reversal boundary between the normal-polarity Brunhes Chron overlying the upper portion of the predominantly reversed polarity Matuyama Chron. Combined with the easily recognized chrons, the Jaramillo and Olduvai normal subchrons within the Matuyama Chron provide a recognizable pattern that forms the basis for paleomagnetic stratigraphy of marine sediments. Although little paleomagnetic work has been performed on Gulf of Mexico sections, the global synchroneity of the reversal pattern makes it possible to establish age assignments on widely separated stratigraphic horizons. This technique has proved instrumental in establishing the ages of (1) important first and last appearances within major fossil groups and (2) Pleistocene oxygen isotope fluctuations. Magnetostratigraphy could have wide applicability in the thick Pleistocene sections that characterize much of the northwestern portions of the Gulf of Mexico. Like paleomagnetic stratigraphy, the hydraulic piston corer (HPC) also could develop into an important stratigraphic and exploration tool in sedimentary basins like the Gulf of Mexico. The main advantage of the HPC is its capacity to recover thick, undisturbed marine sections.

Magnetostratigraphy and the recovery of continuous sedimentary sections through the Deep Sea Drilling Project have improved significantly the stratigraphic resolution of various microfossil groups and have enhanced our understanding of evolutionary events or datums as principal criteria in biochronology. With respect to planktonic foraminiferal populations, evolutionary events have proceeded in the Cenozoic at an average rate of 1.5 new appearances per million years and 1.3 extinctions per million years (Thunell 1981a). The applicability of evolutionary datums in most Pleistocene sections is therefore very difficult, if one considers that the Pleistocene is less than 2 million years in duration. It is thus necessary to subdivide the Pleistocene biostratigraphically by using nonevolutionary events. One of the first techniques applied to Pleistocene sediments of the Gulf Coast region was the application of the zonation of Ericson and Wollin on the basis solely of the presence or absence of *Globorotalia menardii*. This zonation allowed a subdivision on the average of 50,000 years for the latest Pleistocene, and approximately 300,000 years for the remaining Pleistocene. Later work by Kennett and Huddlestun (1972a) used the abundance pattern of *G. menardii* to subdivide the sedimentary sequences of the Gulf of Mexico further.

From the *G. menardii* zonation, a quantitative, total-faunal approach appears to offer the most promise for establishing a high-resolution biostratigraphic scheme for the Gulf of Mexico region. Neff (1983) recently used this technique to divide the Pleistocene into zones of less than 100,000 years in duration. This represents a considerable improvement over the 300,000 year subdivisions of the Ericson and Wollin framework. In addition to biostratigraphic studies, planktonic foraminifera have been important in understanding the paleoclimatic history of the Gulf of Mexico. The newest approach involves a multivariate statistical approach, such as principal components analysis or factor analysis. Foraminiferal assemblages are grouped into temperature-related assemblages and thereby used to estimate water-temperature fluctuations. The application of these quantitative techniques to drilled sections such as the Shell "Eureka" cores and other well samples being released by petroleum companies will provide an unparalleled opportunity for high-resolution studies of the Pleistocene Gulf of Mexico.

Similar opportunities exist for the application of oxygen isotope stratigraphy in high-accumulation-rate sections of the Gulf of Mexico. Oxygen isotope stratigraphy is based on the use of globally synchronous, quasiperiodic fluctuations in the $^{18}O/^{16}O$ ratio of the oceans through time to precisely zone Pleistocene marine sections and correlate these sections within an absolute-time framework with a resolution approaching better than 20,000–50,000 years. Oxygen isotope stages and stage boundaries are well dated through the use of magnetostratigraphy and biostratigraphy. The presence or absence of these isotope stages can be utilized to determine the stratigraphic continuity or presence of hiatuses in various Pleistocene sections. Changes in sedimentation rate can be interpreted in terms of regression and transgression cycles which typified the Pleistocene and played a large role in controlling sedimentation on the continental margin and Mississippi Fan of the Gulf of Mexico. Future oxygen isotope work in the Gulf of Mexico will prove crucial to determining the synchroneity of biostratigraphic datums and individual tephro-chronological events.

The oxygen isotope record in marginal basins like the Gulf of Mexico can also be used to determine past episodes of meltwater runoff from the Laurentide ice sheet. These meltwater events may have played a significant role in controlling deposition rates on the continental margin and in discharging sands from the continental margin into the Mississippi Fan during the Pleis-tocene. In addition, the limited stable isotope data available for the early Pleistocene of the Gulf of Mexico ("Eureka" borehole E67–135 from the DeSoto Canyon) indicates that it is possible to achieve correlations more detailed than the presently available biostratigraphic and lithostratigraphic correlation criteria. The future combination of detailed isotopic and biostratigraphic work holds great promise of significant improvements in Pleistocene stratigraphy of offshore Gulf-of-Mexico sections.

Another powerful stratigraphic tool with great potential in the Gulf of Mexico involves tephrochronology. Owing to its proxim-ity to the volcanically active Middle America region, the Gulf of Mexico is an area of potential deposition of volcanic ash layers. Because volcanic events are geologically instantaneous, ash layers provide an excellent stratigraphic horizon within the sedimentary record. Some of the most widespread volcanic ash layers have occurred during the late Pleistocene. Early work in the tephro-chronology of the Gulf of Mexico dealt with the establishment of a tephra stratigraphy in relationship to foraminiferal biostratig-raphy in piston cores. Ash layers were assigned ages based on the biostratigraphy of each core. For example, the Y/8 ash layer oc-curs during the Y/8 foraminiferal zone established by Kennett and Huddlestun (1972a, b) and shows an age of 84 KYBP. Recent use of the microprobe in tephrochronology allows a precise geochem-ical fingerprinting of tephra layers in terms of the major elemental geochemistry of different eruptions. This technique is highly pre-cise and allows the identification of specific tephra by utilizing small samples. The establishment of a tephrochronologic record with either a detailed biostratigraphy or oxygen isotope stratigraphy makes it possible to extrapolate age assignments of fingerprinted tephra layers to widespread sedimentary sequences. The small sample requirements of the technique make it possible to utilize both dispersed and megascopic tephra horizons, thereby enabling future work to extend a well-dated tephrochronology into both marine or terrestrial sections of the Gulf of Mexico.

Chapter 5 places the four new stratigraphic techniques discussed previously into the context of how the strong climatic variability of the late Cenozoic is preserved in the terrestrial-marine sedimentary record. In particular, how does the continental evidence of glaciation in North America and Europe during the Pleistocene compare with the paleoclimatic record from deep-sea sediments? A synthesis of available terrestrial evidence for glaciation indicates that absolute age determinations of volcanic ash layers in the North American midcontinent fix the maximum age of the Illinoian glaciation and the minimum age of the Kansan glaciation at about 600 KYBP. The maximum age of the Kansan glaciation and minimum age of the Nebraskan glaciation is probably 2.2 MYBP and not 1.2 MYBP as was thought previously. Potassium-argon dates from the Rocky Mountains and Sierra Nevadas support a firm correlation of mountain glaciations with the North American midcontinental glacial record back to about 1.0 MYBP. Recent work in the European Alpine sections highlights a major uncertainty over whether the Biber or the Donau Glacial Events occupied most of the early Matuyama Chron (about 2.5 to 1.5 MYBP). The Biber glaciation is the oldest glacial event in the Alps; thus, if the Biber dates from the early Matuyama Chron, the first European mountain glaciation would have postdated the Nebraskan and Deadman Pass glacial events of North America by about 1.0 MYBP. The Cromer Till or Elbe glaciation of the northern-European lowland is probably correlative with the late Kansan (< 1.0 MYBP) of North America. Elbe deposits constitute the oldest evidence of lowland glaciation found to date in Europe. Loess and pollen stratigraphies extend the European record of periglacial conditions back to about 1.7 MYBP and 2.5 MYBP, respectively, suggesting that early, pre-Elbe lowland glacial deposits may remain undiscovered.

Deep-sea records of temperature change, including paleoclimatic inferences drawn from the Ericson-Wollin biostratigraphic zones, appear to be out-of-phase with terrestrial glacial events. Relatively warm seas correlate with continental glacial periods, but the two sets of records demonstrate similar periodicities. In contrast to oceanic records of temperature change, the record of sea-level variations in continental-margin sediments of the northern Gulf of Mexico seem to have an in-phase relationship with terrestrial glacial events. Deep-sea oxygen isotope records, which reflect variations in global ice volume, vary at a much higher frequency than has so far been observed in most continental glacial and sea-level records. Longer-period trends in isotopic variation do not appear to correlate well either with terrestrial glacial records or oceanic paleotemperature records. The apparent divergence of temperature, sea-level, and oxygen isotope records during the Pleistocene may be explained by a two–ice-sheet model in the Northern Hemisphere as follows. A midlatitude continental ice mass and a floating Arctic Ocean ice mass vary somewhat independently in terms of volume. The basis of this model is the control which insolation (solar radiation) exerts on the distribution of heat on the Earth's surface by periodic, and therefore predictable, variations in orbital parameters. Midlatitude insolation controls midcontinental ice sheets, which thereby vary primarily in response to 23,000 year and longer orbital periodicities. High-latitude insolation varies primarily in response to 41,000 year and longer orbital periodicities, thus exerting primary control on the extent of ice in the Arctic Ocean.

The potential of obtaining new biostratigraphic and oxygen-isotopic data from high-accumulation-rate Pleistocene sections of the Gulf of Mexico will enhance our understanding of how sea-level and climatic change have controlled the sites and rates of terrigenous-clastic deposition, both along the continental margin and on deep-water depositional features like the Mississippi Fan. This basic information will, in turn, enable paleoclimatic models to be constructed to test why apparent differences exist between sea level, open-ocean paleotemperatures, and terrestrial glacial records. Another concluding point is the suggestion that although the techniques discussed herein have emphasized the Pleistocene of the Gulf of Mexico, they each have applicability to other sedimentary basins and other time periods. Work in progress in several laboratories will test this potential in the near future.

References

Agassiz, Louis. 1840. *Études sur les glaciers*. Neuchâtel: Switzerland (privately published).

Aharon, P. 1983. 140,000-yr isotope climatic record from raised coral reef in New Guinea. *Nature*. 304:720–723.

Akers, W. H. 1965. Pliocene-Pleistocene boundary, northern Gulf of Mexico. *Science*. 149:741–742.

Akers, W. H. and Holk, A. J. 1957. Pleistocene beds near the edge of the continental shelf, southeastern Louisiana. *Geological Society of America Bulletin*. 68:983–992.

Andrews, J. T.; Shilts, W. W. and Miller, G. H. 1983. Multiple deglaciations of the Hudson Bay Lowlands, Canada, since deposition of the Missinaibi (last glacial?) Formation. *Quaternary Research*. 19:18–37.

Backman, J. 1979. Pliocene biostratigraphy of DSDP Sites 111 and 116 from the North Atlantic Ocean and the age of Northern Hemisphere glaciation. *Stockholm Contribution in Geology*. 33:115–137.

Backman, J.; Shackleton, N. J.; and Tauxe, L. 1983. Quantitative nannofossil correlation to open ocean deep-sea sections from Plio-Pleistocene boundary at Vrica, Italy. *Nature*. 304:156–158.

Baker, P. A.; Gieskes, J. M.; and Elderfield, H. 1982. Diagenesis of carbonate in deep-sea sediments—evidence from Sr/Ca ratios and interstitial-dissolved Sr^{2+} data. *Journal of Sedimentary Petrology*. 52:71–82.

Bandy, O. L. 1960. The geologic significance of coiling ratios in the foraminifera *Globigerina pachyderma* (Ehrenberg). *Journal of Paleontology*. 34:671–681.

Bé, A. W. H. 1970. *Globorotalia menardii* flexuosa (Koch): An "extinct" foraminiferal subspecies living in the northern Indian Ocean. *Deep-Sea Research* 17:595–501.

Bé, A. W. H. 1977. An ecological, zoogeographic and taxonomic review of Recent planktonic foraminifera. In: *Oceanic Micropaleontology*, v. 1. Ramsay, A. T. S., ed. 1–100. New York: Academic Press.

Beard, J. H. 1969. Pleistocene paleotemperature record based on planktonic foraminifers, Gulf of Mexico. *Gulf Coast Association of Geological Societies Transactions*. 19:535–553.

———— 1973. Pleistocene-Holocene boundary, Wisconsinan substages, Gulf of Mexico. *Geological Society of America Memoir 136*. 277–316.

Beard, J. and Lamb, J. L. 1968. The lower limit of the Pliocene and Pleistocene in the Caribbean and Gulf of Mexico: *Gulf Coast Association of Geological Societies Transactions*. 18:174–186.

Beard, J. H.; Sangree, J. B.; and Smith, L. A. 1982. Quaternary chronology, paleoclimate, depositional sequences and eustatic cycles. *American Association of Petroleum Geologists Bulletin*. 66:158–169.

Berger, A. L. 1976. Obliquity and precession for the last 5,000,000 years. *Astronomy and Astrophysics*. 5:127–135.

———— 1978. Long-term variations of caloric insolation resulting from the Earth's orbital elements. *Quaternary Research*. 9:139–167.

Berger, W. H. 1977. Carbon dioxide excursions and the deep sea record: aspects of the problem. In: *The Fate of Fossil Fuel CO_2 in the Oceans*. Andersen, N. R. and Malahoff, A., eds. 505–542. New York: Plenum Press.

———— 1979. Stable isotopes in Foraminifera. In: *Foraminiferal Ecology and Paleoecology*. Society of Economic Paleontologists and Mineralogists Short Course No. 6. Houston: Society of Economic Paleontologists and Mineralogists.

———— 1977b. Global freshening of the upper ocean layer during deglaciation. *Naturwissenschaften*. 64:634.

———— 1982. On the definition of the Pleistocene/Holocene boundary in deep-sea sediments. *Sveriges Geologiska Undersokning, Avhand. Uppsat.* 76:270–280.

Berggren, W. A. 1972. Late Pliocene-Pleistocene glaciation. In: *Initial Reports of the Deep Sea Drilling Project.* vol. 22. Laughton, A. S.; Berggren, W. A. et al., eds. 953–963. Washington, D.C.: U.S. Government Printing Office.

———— 1973. The Pliocene time scale: Calibration of planktonic foraminiferal and calcareous nannoplankton zones. *Nature.* 243:391–397.

———— 1977. The Pliocene/Pleistocene boundary in Deep-Sea Sediments: Status in 1975, *Giornale di Geologia* 41 (2):375–384.

Berggren, W. A. and Van Couvering, J. A. 1974. The Late Neogene: Biostratigraphy, geochronology and paleoclimatology of the last 15 million years in marine and continental sequences. *Paleogeography, Paleoclimatology, Paleoecology.* 16:1–216.

Berggren, W. A.; Burckle, L. H.; Cita, M. B.; Cooke, H. B. S.; Funnell, B. M.; Gartner, S.; Hays, J. D.; Kennett, J. P.; Opdyke, N. D.; Pastouret, L.; Shackleton, N. J.; and Takayanagi, Y. 1980. Towards a Quaternary Time Scale. *Quaternary Research.* 13:277–302.

Berggren, W. A.; Van Couvering, J. A.; and Kent, D. V. 1983. The Neogene. In: *Geochronology and the Geologic Record.* Snelling, N. J., ed. Geological Society of London, Special Paper. (In press.)

Bernard, H. A. 1950. Quaternary geology of southeast Texas. Ph.D. dissertation. Louisiana State University. Baton Rouge.

Bernard, H. A. and LeBlanc, R. G. 1965. Résumé of the Quaternary geology of the northwestern Gulf of Mexico provence. In: *The Quaternary of the United States.* Wright, H. P., Jr., and Frey, D. G., eds. 137–185. Princeton: Princeton University Press.

Birkeland, P. W.; Crandell, D. R.; and Richmond, G. M. 1971. Status of correlation of Quaternary stratigraphical units in the western conterminous United States. *Quaternary Research.* 1:208–227.

Blackwelder, B. W.; Pilkey, O. H.; and Howard, J. D. 1979. Late-Wisconsinan sea levels on the southeast U.S. Atlantic Shelf based on in-place shoreline indicators. *Science.* 204:618–620.

Bloom, A. M.; Broecker, W. S.; Chappell, J. M. A.; Matthews, R. K.; and Mesolella, K. J. 1974. Quaternary sea level fluctuations on a tectonic coast: New $^{230}Th/^{234}U$ dates from the Huon Peninsula, New Guinea. *Quaternary Research.* 4:185–205.

Boellstorff, J. 1978. North American Pleistocene stages reconsidered in light of probable Pliocene-Pleistocene continental glaciation. *Science.* 202:305–307.

Bolli, H. M. and Premoli, Silva, I. 1973. Oligocene to Recent planktonic foraminifera and stratigraphy of the Leg 15 sites in the Caribbean Sea. In: *Initial Reports of the Deep Sea Drilling Project* v. 15. Edgar, N. R.; Saunders, J. B.; et al., eds. 475–498. Washington, D.C.: U.S. Government Printing Office.

Borchardt, G. A.; Harwood, M. E.; and Schmitt, R. A. 1971. Correlation of volcanic-ash deposits by activation analysis of glass separates. *Quaternary Research.* 1:247–260.

Bouma, A. H.; Smith, L. B.; Sidner, B. R.; and McKee, T. R. 1978. Intraslope basin in northwest Gulf of Mexico. In: *Framework, Facies and Oil-Trapping Characteristics of the Upper Continental Margin.* Bouma, A. H.; Moore, G. T.; and Coleman, J. M., eds. American Association of Petroleum Geologists Studies in Geology. No. 7. Tulsa: American Association of Petroleum Geologists.

Bowles, F. A.; Jack, R. N.; and Carmichael, I. S. E. 1973. Investigation of deep-sea volcanic ash layers from equatorial Pacific cores. *Geological Society of America Bulletin.* 84:2371–2388.

Bramlette, M. N. and Bradley, W. H. 1941. Geology and biology of North

Atlantic deep sea cores between Newfoundland and Ireland, Part I. Lithology and geologic interpretation. *U.S. Geological Survey Professional Paper.* 196-A: 1–55.

Bray, J. R., 1977. Pleistocene volcanism and glacial initiation. *Science.* 197:251–254.

―――― 1979. Neogene explosive volcanicity, temperature, and glaciation. *Nature.* 282:603–605.

Briskin, M. and Berggren, W. A. 1975. Pleistocene stratigraphy and quantitative paleoceanography of tropical North Atlantic core V16–205. In: *Late Neogene Epoch Boundaries.* Saito, T. and Burkle, L. H., eds. 167–198. Micropaleontology Press Special Publication No. 1. New York: American Museum of Natural History.

Briskin, M. and Harrell, J. 1980. Time-series analysis of the Pleistocene deep-sea paleoclimatic record. *Marine Geology.* 36:1–22.

Broecker, W. S. 1965. Isotope geochemistry and the Pleistocene climatic record. In: *The Quaternary of the United States.* Wright, H. E., Jr., and Frey, D. G., eds. 737–753. Princeton: Princeton University Press.

―――― 1966. Absolute dating and the astronomical theory of glaciation. *Science.* 151:299–302.

―――― 1975. Floating glacial ice caps in the Arctic Ocean. *Science.* 188:1116–1118.

Broecker, W. S. and Ku, T. L. 1969. Caribbean cores P6304-8 and P6304-9: New analysis of absolute chronology. *Science.* 166:404–406.

Broecker, W. S. and Van Donk, J. 1970. Insolation changes, ice volumes, and the ^{18}O record in deep-sea cores. *Review of Geophysics and Space Physics.* 8:169–198.

Broecker, W. S.; Thurber, D. L.; Goddard, J.; Ku, T. L.; Matthews, R. K.; and Mesolella, K. J. 1968. Milankovitch hypothesis supported by precise dating of coral reefs and deep-sea sediments. *Science.* 159:297–301.

Brunnaker, K.; Löscher, M.; Tillmans, W.; and Urban, B. 1982. Correlation of the Quaternary terrace sequence in the lower Rhine Valley and northern Alpine Foothills of Central Europe. *Quaternary Research.* 18:152–173.

Brunner, C. A. 1979. Distribution of planktonic foraminifera in surface sediments of the Gulf of Mexico. *Micropaleontology.* 25:325–335.

―――― 1982. Paleoceanography of surface waters in the Gulf of Mexico during the Late Quaternary. *Quaternary Research.* 17:105–119.

Brunner, C. A. and Cooley, J. F. 1976. Circulation in the Gulf of Mexico during the late glacial maximum, 18,000 years ago. *Geological Society of America Bulletin.* 87:681–686.

Brunner, C. A. and Keigwin, L. D. 1981. Late Neogene biostratigraphy and stable isotope stratigraphy of a drilled core from the Gulf of Mexico. *Marine Micropaleontology.* 6:397–418.

Burckle, L. H.; Keigwin, L. D., Jr.; Opdyke, N. D.; and Theyer, T. 1979. Global Miocene correlation using stable isotopes, microfossils, magnetostratigraphy and lithology. *Geological Society of America Abstracts with Programs.* 11:396.

Capurro, L. R. A. and Reid, J. L., eds. 1972. *Contributions on the Physical and Geophysical Oceanography of the Gulf of Mexico.* Texas A & M University Oceanographic Studies Vol. 3. Houston: Gulf Publishing Company.

Chamberlin, T. C. 1895. The classification of American glacial deposits. *Journal of Geology.* 3:270–277.

―――― 1896. Nomenclature of glacial formations. *Journal of Geology.* 4:872–876.

Chapin, C. E. and Elston, W. E. 1979. Ash-flow tuffs. *Geological Society of America Special Paper.* No. 180. Boulder: Geological Society of America.

Chen, M.-P. 1978. Calcareous nannoplankton biostratigraphy and paleoclimatic history of the late Neogene sediments of the Northwest Florida Continental Shelf. Ph.D. dissertation. Texas A & M University. College Station.

Christiansen, R. L.; Obradovich, J. D.; and Polank, H. R., Jr. 1968. Late Cenozoic volcanic stratigraphy of the Yellowstone Park region—a preliminary report (abstract). *Geological Society of America Special Paper*. 121:591–592.

Cita, M. B.; Vergnaud-Grazzini, C.; Robert, C.; Chamley, H.; Ciaranfi, N.; and d'Onofrio, S. 1977. Paleoclimatic record of a long deep sea core from the eastern Mediterranean. *Quaternary Research*. 8:205–235.

CLIMAP Project Members. 1976. The surface of the Ice-Age Earth. *Science*. 191:1131–1137.

CLIMAP, 1981. Seasonal reconstructions of the Earth's surface at the last glacial maximum. *Geological Society of America Map and Chart Series MC-36*. Boulder, Colorado: Geological Society of America.

Colalongo, M. L.; Pisini, G.; Pelosio, G.; Raffi, S.; Rio, D.; Ruggieri, G.; Sartoni, S.; Selli, R. and Sprovieri, R. 1982. The Neogene/Quaternary Boundary Definition: A review and proposal. *Geogr. Fis. Dinam. Quat.* 5:59–68.

Cooke, H. B. S. 1972. Pleistocene chronology: Long or short? *Maritime Sediments*. 8:1–12.

Cox, A. 1969. Geomagnetic reversals. *Science*. 163:237–245.

Cox, A.; Doell, R. R.; and Dalrymple, G. B. 1965. Quaternary paleomagnetic stratigraphy. In: *The Quaternary of the United States*. Wright, H. E., Jr. and Frey, D. G., eds. International Association of Quaternary Research 7th Congress. Princeton: Princeton University Press.

Craig, H. 1957. Isotopic standards for carbon and oxygen and correction factors for mass-spectrometric analysis of carbon dioxide. *Geochimica et Cosmochimica Acta*. 12:133–149.

Craig, H. and Gordon, L. I. 1965. Deuterium and oxygen-18 variations in the ocean and the marine atmosphere. In: *Stable Isotopes in Oceanographic Studies and Paleotemperatures, Spoleto 1965*. Tongiori, E., ed. 9–130. Pisa: Consiglio Nazionale delle Ricerche, Labaratorio di Geologia Nucleare.

Curry, R. P. 1966. Glaciation about 3,000,000 years ago in the Sierra Nevada. *Science*. 154:770–771.

Dansgaard, W. and Tauber, H. 1969. Glacier oxygen-18 content and Pleistocene ocean temperatures. *Science*. 166:499–502.

Davis, J. C. 1973. *Statistics and Data Analysis in Geology*. New York: John Wiley and Sons.

Deines, P. 1970. Mass spectrometer correction factors for the determination of small isotopic composition variations of carbon and oxygen. *International Journal of Mass Spectrometry and Ion Physics*. 4:283.

Deuser, W. G.; Ross, E. H.; and Waterman, L. S. 1976. Glacial and pluvial periods: Their relationship revealed by Pleistocene sediments of the Red Sea and Gulf of Aden. *Science*. 191:1168–1170.

Dodge, R. E.; Fairbanks, R. G.; Benninger, L. K.; and Maurrasse, R. 1983. Pleistocene sea levels from raised coral reefs of Haiti. *Science*. 219:1423–1425.

Drexler, J. W.; Rose, W. I., Jr.; Sparks, R. S. J.; and Ledbetter, M. T. 1980. The Los Chocoyos Ash, Guatemala: a major stratigraphic marker in Middle America and in three ocean basins. *Quaternary Research*. 13:327–345.

Duplessy, J.-C. 1978. Isotope studies. In: *Climatic Change*. Griffin, J., ed. 46–67. Cambridge: Cambridge University Press.

Eberl, B. 1930. *Die Eiszeitenfolge im Nordlichen Alpenvorlande*. Ausburg (privately published).

Elderfield, H.; Gieskes, J. M.; Baker, P. A.; Oldfield, R. K.; Hawkesworth, C. J.; and Miller, R. 1982. $^{87}Sr/^{86}Sr$ and $^{18}O/^{16}O$ ratios, interstitial water chemistry and diagenesis in deep-sea carbonate sediments on the Ontong-Java Plateau. *Geochimica et Cosmochimica Acta.* 64:2259–2268.

Emiliani, C. 1955. Pleistocene temperatures. *Journal of Geology.* 63:538–578.

—— 1958. Paleotemperature analysis of core 280 and Pleistocene correlations. *Journal of Geology.* 66:264–275.

—— 1961. Cenozoic climatic changes as indicated by the stratigraphy and chronology of deep-sea cores of *Globigerina* ooze facies. *Annals of the New York Academy of Science.* 95:521.

—— 1964. Paleotemperature analysis of the Caribbean cores A254-BR-C and CP-28. *Geological Society of America Bulletin.* 75:129–144.

—— 1966. Paleotemperature analysis of the Caribbean cores P6304–8 and P6304–9 and a generalized temperature curve for the last 425,000 years. *Journal of Geology.* 74:109–126.

—— 1971a. Isotopic paleotemperatures and shell morphology of *Globigerinoides ruber* in the type section for the Pliocene/Pleistocene boundary. *Micropaleontology.* 17:233–238.

—— 1971b. The amplitude of Pleistocene climatic cycles at low latitudes and the isotopic composition of glacial ice. In: *The Late Cenozoic Glacial Ages.* Turekian, K. K., ed. 183–187. New Haven: Yale University Press.

—— 1972. Quaternary paleotemperatures and the duration of the high-temperature intervals. *Science.* 178:398–401.

—— 1978. The cause of the ice ages. *Earth and Planetary Science Letters.* 37:349–352.

Emiliani, C. and Rona, E. 1969. Caribbean cores P6304–8 and P6304–9: New analysis of absolute chronology, A reply. *Science.* 166:1551.

Emiliani, C.; Mayeda, T.; and Selli, R. 1961. Paleotemperature analysis of the Plio-Pleistocene sections at Le Costella, Calabria, Southern Italy. *Geological Society of America Bulletin.* 72(5):679–688.

Emiliani, C.; Gartner, S.; Lidz, B.; Eldridge, K.; Elvey, D. K.; Huang, T. C.; Stipp, J. J.; and Swanson, M. F. 1975. Paleoclimatological analysis of Late Quaternary cores from the northeastern Gulf of Mexico. *Science.* 189:1083–1088.

Epstein, S. and Lowenstam, H. A. 1953. Temperature shell-growth relations of Recent and interglacial Pleistocene shoal water biota from Bermuda. *Journal of Geology.* 61:424–438.

Epstein, S. and Mayeda, T. 1953. Variation of ^{18}O content of waters from natural sources. *Geochimica et Cosmochimica Acta.* 4:213–224.

Ericson, D. B. 1959. Coiling direction of *Globigerina pachyderma* as a climatic index. *Science.* 130:219–220.

Ericson, D. B. and Wollin, G. 1956a. Correlation of six cores from the equatorial Atlantic and the Caribbean. *Deep-Sea Research.* 3:104–125.

—— 1956b. Micropaleontological and isotopic determinations of Pleistocene climates. *Micropaleontology.* 2:257–270.

—— 1968. Pleistocene climates and chronology in deep-sea sediments. *Science.* 162:1227–1234.

—— 1970. Pleistocene climates in the Atlantic and Pacific Oceans: A comparison based on deep-sea sediments. *Science.* 167:1483–1485.

Ericson, D. B.; Wollin, G.; and Wollin, J. 1954. Coiling direction of *Globorotalia truncatulinoides* in deep-sea cores. *Deep-Sea Research.* 2:152–158.

Ericson, D. B.; Ewing, M.; Wollin, G.; and Heezen, B. C. 1961. Atlantic deep-sea sediment cores. *Geological Society of America Bulletin.* 72:193.

Ericson, D. B.; Ewing, W. M.; and Wollin, G. 1964. The Pleistocene Epoch in deep-sea sediments. *Science.* 146:723–732.

Ewing, M.; Ericson, D. B.; and Heezen, B. C. 1958. Sediments and topography of the Gulf of Mexico. In: *Habitat of Oil.* Weeks, L. G., ed. 995–1053. Tulsa: American Association of Petroleum Geologists.

Ewing, M.; Heezen, B. C. and Ericson, D. 1959. Significance of the Worzel Ash. *U.S. National Academy of Science Proceedings.* 45:355–361.

Ewing, J. I.; Worzel, J. L.; and Ewing, M. 1962. Sediments and oceanic structural history of the Gulf of Mexico. *Journal of Geophysical Research* 67:2509–2527.

Fairbanks, R. G. and Matthews, R. K. 1978. The marine oxygen isotope record in Pleistocene coral, Barbados, West Indies. *Quaternary Research.* 10:181–196.

Falls, W. F. 1980. Glacial meltwater inflow into the Gulf of Mexico during the last 150,000 years: Implications for isotope stratigraphy and sea level studies. M.S. Thesis, University of South Carolina, Columbia.

Federman, A. N. and Carey, S. N. 1980. Electron microprobe correlation of tephra layers from eastern Mediterranean abyssal sediments and the island of Santorini, *Quaternary Research.* 13:160–171.

Field, M. E.; Meisburger, E. P.; Stanley, E. A.; and Williams, S. J. 1979. Upper Quaternary deposits on the Atlantic inner shelf of the United States. *Geological Society of America Bulletin.* 90:618–629.

Fillon, R. H. 1977. Ice-rafted detritus and paleotemperature: Late Cenozoic relationships in the Ross Sea region. *Marine Geology.* 25:73–93.

Fillon, R. H. and Williams, D. F. 1983. Glacial evolution of the Plio-Pleistocene: role of continental and Arctic Ocean ice sheets, *Paleogeography, Paleoclimatology, Paleoecology.* 42:7–33.

Fillon, R. H.; Miller, G. H.; and Andrews, J. T. 1981. Terrigenous sand in Labrador Sea hemipelagic sediments and paleoglacial events on Baffin Island over the last 100,000 years. *Boreas.* 10:107–124.

Fisk, H. N. 1939. Depositional terrace slopes in Louisiana. *Journal of Geomorphology.* 2:181–200.

———— 1944. *Geological Investigation of the Alluvial Valley of the Lower Mississippi River.* Vicksburg: Mississippi River Commission.

Foote, R. Q.; Martin, R. G.; Powers, R. B. 1983. Oil and gas potential of the Maritime Boundary region of the central Gulf of Mexico. *American Association of Petroleum Geologists.* 67:1047–1065.

Forbes, E. 1846. On the connection between the distribution of the existing fauna and flora of the British Isles and the geographical changes which have affected their area, especially during the epoch of the Northern Drift. *Great Britain Geological Survey Memoir.* 1:336–432.

Frye, J. C. and Willman, H. B. 1960. Classification of the Wisconsinan Stage in the Lake Michigan glacial lobe. *Illinois Geological Survey Circular.* 285:1–16.

Frye, J. C.; Willman, H. B.; and Black, R. F. 1965. Outline of glacial geology of Illinois and Wisconsin. In: *The Quaternary of the United States.* Wright, H. E., Jr. and Frey, D. G., eds. 43–61. International Association of Quaternary Research 7th Congress. Princeton: Princeton University Press.

Gignoux, M. 1910. Sur la classification du Pliocene et du Quaternaire de l'Italie du Sud. *Centre de Recherche d'Académe de Science Paris* 150, 841–844.

———— 1913. Les formations marines Pliocenes et Quaternaires de l'talie du Sud et de la Sicile. *Annals Université de Lyon.* 36:1–693.

Haggerty, S. E. 1970. Magnetic minerals in pelagic sediments, Carnegie Inst. Wash. Year Book, 68:332–336.

Hahn, G. A.; Rose, W. I., Jr.; and Meyers, T. 1979. Geochemical correlation of genetically related rhyolitic ash flow and air fall ashes, central and western Quatemala and the equatorial Pacific. In: *Ash Flow Tuffs.* Chapin, C. E. and

Elston, W. E., eds. 101–112. Geological Society of America Special Paper. 180.

Haq, B. U.; Berggren, W. A. and Van Couvering, J. A. 1977. Corrected age of the Pliocene/Pleistocene boundary. *Nature.* 269:483–488.

Hardenbol, J. and Berggren, W. A. 1978. A new Paleogene numerical time scale. In: *Contributions to the Geological Time Scale.* Cohee, G.; Glaessner, M. and Hedberg, H., eds. 213–234. American Association of Petroleum Geologists Studies in Geology No. 6.

Harrison, C. G. A. 1966. The paleomagnetism of deep-sea sediments. *J. Geophys. Res.,* 71:3033–3043.

Harrison, C. G. A. and Funnell, B. M. 1964. Relationship of paleomagnetic reversals and micropaleontology in two late Cenozoic cores from the Pacific Ocean. *Nature,* 204:566–567.

Hays, J. D. and Shackleton, N. D. 1976. Globally synchronous extinction of the radiolarian *Stylotractus universus. Geology.* 4:649–652.

Hays, J. D.; Saito, T.; Opdyke, N. D.; and Burckle, L. H. 1969. Pliocene-Pleistocene sediments of the equatorial Pacific: Their paleomagnetic, biostratigraphic and climatic record. *Geological Society of America Bulletin.* 80:1481–1514.

Hays, J. D. and Opdyke, N. D. 1967. Antarctic radiolaria, magnetic reversals, and climatic change. *Science.* 158:1001–1010.

Hays, J. D.; Imbrie, J.; and Shackleton, N. J. 1976. Variations in the Earth's orbit: Pacemaker of the ice ages. *Science.* 194:1121–1132.

Hecht, A. 1976. The oxygen isotope record of foraminifera in deep-sea sediments. In: *Foraminifera* Vol. 2. Hedly, R. H. and Adams, C. G. eds. 1–43. London: Academic Press.

Hedberg, H. D. 1976. International Stratigraphic Guide—a guide to stratigraphic classification, terminology and procedure. New York: John Wiley and Sons.

Hein, J. R.; Scholl, D. W.; and Miller, J. 1978. Episodes of Aleutian Ridge explosive volcanism. *Science.* 199:137–141.

Heirtzler, J. R.; Dickson, G. O.; Herron, E. N.; Pitman, W. C. and LePichon, X. 1968. Marine magnetic anomalies, geomagnetic field reversals and motions of the ocean floor and continents. *J. Geophys. Res.* 73:2119–2136.

Herman, Y. and Hopkins, D. 1980. Arctic oceanic climate in late-Cenozoic time. *Science* 209:557–562.

Holland, D. S.; Nunan, W. E.; Lammelin, D. R.; and Woodhams, R. L. 1980. Eugene Island Block 330 field, offshore Louisiana. In: *Grant Oil and Gas Fields of the Decade: 1968–1978.* Halbouty, M. T., ed. 253–280. American Association of Petroleum Geologists Memoir.

Howell, K. L. 1982. Paleoclimatic and paleoceanographic interpretations of stable isotopic results from DSDP Sites 516A and 517. M.S. thesis, University of South Carolina, Columbia, S.C.

Huang, T. C.; Watkins, N. D.; Shaw, D. M.; and Kennett, J. P. 1973. Atmospherically transported volcanic dust in South Pacific deep-sea sedimentary cores at distances over 3000 km from the eruptive source. *Earth and Planetary Science Letters.* 20:119–124.

Huang, T. C.; Watkins, N. D.; and Shaw, D. M. 1975. Atmospherically transported volcanic glass in deep-sea sediments: Development of a separation and counting technique. *Deep-Sea Research.* 22:185–196.

Hubbard, C. W.; Ray, D. E.; Savage, D. E.; Taylor, D. W.; and Guilday, J. E. 1965. Quaternary mammals of North America, In: *The Quaternary of the United*

States. Wright, Jr. H. E. and Frey, D. G., eds. 509–525. International Association for Quaternary Research. 7th Congress. Princeton: Princeton University Press.

Hughes, T.; Denton, G. H.; and Grosswald, M. G. 1977. Was there a late-Würm ice sheet? *Nature:* 266:596–602.

Imbrie, J. 1963. Factor and vector analysis programs for analyzing geologic data. *Technical Report No. 6.* Office of Naval Research. Washington, D.C.: U.S. Government Printing Office.

Imbrie, J. and Kipp, N. G. 1971. A new micropaleontological method for quantitative paleoclimatology: application to a late Pleistocene Caribbean core. In: *The Late Cenozoic Glacial Ages.* Turekian, K. K., ed. 71–181. New Haven: Yale University Press.

Imbrie, J.; Van Donk, J.; and Kipp, N. G. 1973. Paleoclimatic investigation of a late Pleistocene Caribbean deep-sea core: Comparison of isotopic and faunal methods. *Quaternary Research.* 3:10–38.

Imbrie, J.; Hays, J. D.; Martinson, D. G.; McIntyre, A.; Mix, A. C.; Morley J. J.; Pisias, N. G.; Prell, W. L.; and Shackleton, N. J. In press. The orbital theory of Pleistocene climate: support for a revised chronology of the marine $\delta^{18}O$ record. In: *Milankovitch and Climate.* Berger, A. et al., ed. Boston: D. Riedel Publ.

Ingle, J. C., Jr. 1967. Foraminiferal biofacies variation and the Miocene-Pliocene Boundary in California. *Bulletin of American Paleontology.* 52:210–394.

———— 1973. Summary comments on Neogene biostratigraphy, physical stratigraphy and paleoceanography in the marginal northeastern Pacific Ocean. In: *Initial Reports of the Deep Sea Drilling Project.* Vol. 18. Rulm, L. D.; von Huene, R.; et al., eds. 949–960. Washington, D.C.: U.S. Government Printing Office.

Izett, G. A.; Wilcox, R. E.; Obradovich, J. D.; and Reynolds, R. L. 1971. Evidence for two Pearlette-like ash beds in Nebraska and adjoining areas. *Geological Society of America Abstracts with Programs.* 7:610.

Jenkins, D. G. 1967. Planktonic foraminiferal zones and new taxa from the Pleistocene of New Zealand. *New Zealand Journal of Geology and Geophysics.* 10:1064–1078.

Keigwin, L. D., Jr. 1979. Late-Cenozoic stable-isotope stratigraphy and paleoceanography of DSDP Sites from the east equatorial and north central Pacific Ocean. *Earth and Planetary Science Letters.* 45:361–382.

———— 1982. Appendix: Basis for age assignments at Deep Sea Drilling Project Sites 502 and 503. In: *Initial Reports of the Deep-Sea Drilling Project.* Vol. 68. Prell, W. L.; Gardner, J. V.; et al., eds. 493–495. Washington, D.C.: U.S. Government Printing Office.

———— 1982. Stable isotope stratigraphy and paleoceanography of Sites 502 and 503. In: *Initial Reports of the Deep-Sea Drilling Project.* Vol. 68. Prell, W. L.; Gardner, J. V. and others, eds. 445–453. Washington, D.C.: U.S. Government Printing Office.

Keller, J.; Ryan, W. B. F.; Ninkovich, D.; and Altherr, R. 1978. Explosive volcanic activity in the Mediterranean over the past 200,000 yr as recorded in deep-sea sediments. *Geological Society of America Bulletin.* 89:591–604.

Kennett, J. P. 1976. Phenotypic variation in some Recent and Late Cenozoic planktonic foraminifera. In: *Foraminifera.* Vol. 2. Hedley, R. H. and Adams, C. G., eds. 111–169. London: Academic Press.

———— 1981. Marine tephrochronology, In: *The Sea.* Vol. 7. *The Oceanic Lithosphere.* Emiliani, C., ed. New York: John Wiley and Sons.

Kennett, J. P. and Huddleston, P. 1972a. Late-Pleistocene paleoclimatology, foraminiferal biostratigraphy, and tephrochronology, Western Gulf of Mexico. *Quaternary Research.* 2:38–69.

———— 1972b. Abrupt climatic change at 90,000 YBP: faunal evidence from Gulf of Mexico cores. *Quaternary Research.* 2:384–395.

Kennett, J. P. and Watkins, N. D. 1970. Geomagnetic polarity change, volcanic maxima, and faunal extinction in the South Pacific. *Nature.* 277:930–934.

Kennett, J. P. and Shackleton, N. J. 1975. Laurentide ice sheet meltwater recorded in Gulf of Mexico deep-sea cores. *Science.* 188:147–150.

Kennett, J. P. and Thunell, R. C. 1975. Global increase in Quaternary explosive volcanism. *Science.* 187:497–503.

———— 1977. On explosive Cenozoic volcanism and climatic implications. *Science.* 196:1231–1234.

Kennett, J. P.; McBirney, A. R. and Thunell, R. C., 1977. Episodes of Cenozoic volcanism in the circum-Pacific region. *Journal of Volcanology and Geothermal Research.* 2:145–163.

Kent, D.; Opdyke, N. D.; and Ewing, M. 1971. Climate change in the North Pacific using ice-rafted detritus as a climatic indicator. *Geological Society of America Bulletin.* 82:2741–2745.

Kent, D. V. and Lowrie, W. 1974. Origin of magnetic instability in sediment cores from the central North Pacific. *J. Geophys. Res.* 79:2987–3000.

Kent, D. V. and Spariosu, D. J. 1982. Magnetostratigraphy of Caribbean Site 502 hydraulic piston cores. In: *Initial Reports of the Deep-Sea Drilling Project.* Vol. 68: 419–434. Prell, W. L.; Gardner, J. V.; et al., eds. Washington, D.C.: U.S. Government Printing Office.

Keunen, P. H. and Neeb, G. A. 1943. Geological results. In: *The Snellius Expedition.* Vol. 5, Part 3. Leyden: Brill.

Kullenberg, B., 1947. The piston core samples. *Svenska Hydrog.-Biol. Komm. Skr.* 3:1–46.

Killingley, J. S. 1983. Effects of diagenetic recrystallization on the $^{18}O/^{16}O$ values of deep-sea sediments. *Nature.* 301:594–597.

Klovan, J. E. and Imbrie, J. 1971. An algorithm and FORTRAN-IV program for large scale Q-mode factor analysis and calculation of factor scores. *Mathematical Geology.* 3:61–77.

Klovan, J. E. and Miesch, A. T. 1976. EXTENDED CABFAC and QMODEL computer programs for Q-model factor analysis of compositional data. *Computers and Geoscience* L:161–178.

Kominz, M. A.; Heath, G. R.; Ku, T. L.; and Pisias, N. G. 1979. Brunhes time scales and the interpretation of climatic change. *Earth and Planetary Science Letters.* 45:394–410.

Ku, T. L. 1966. Uranium series disequilibrium in deep-sea sediments. Ph.D. dissertation. Columbia University, New York, N.Y.

Kukla, G. J. 1977. Pleistocene land-sea correlations: 1. Europe. *Earth Science Reviews.* 13:307–374.

LaBrecque, J. L.; Kent, D. V.; and Cande, S. C. 1977. Revised magnetic polarity time scale for late Cretaceous and Cenozoic time. *Geology.* 5:330–335.

Lamb, J. L. 1969. Planktonic foraminiferal datums and late Neogene epoch boundaries in the Mediterranean Caribbean and Gulf of Mexico. *Gulf Coast Association of Geological Societies Transactions.* 19:559–578.

Lamb, J. and Beard, J. D. 1972. Late Neogene planktonic foraminifera in the Caribbean, Gulf of Mexico and Italian stratotypes. *University of Kansas Paleontological Contribution Article 57.* 1–67. Lawrence, Kansas: University of Kansas.

Ledbetter, M. T. 1982. Tephrochronology at Sites 502 and 503. In: *Initial Reports of Deep Sea Drilling Project.* Vol. 68. Prell, W. and Gardner, J. D. et al., eds. 403–408. Washington, D.C.: U.S. Government Printing Office.

Ledbetter, M. T. and Ciesielski, P. F. 1982. Bottom-current erosion along a traverse in the South Atlantic sector of the Southern Ocean. *Marine Geology.* 46:329–342.

Leighton, M. M. 1968. The Iowan glacial drift sheets of Iowa and Illinois. *Journal of Geology.* 76:259–279.

Leighton, M. M. and Willman, H. B. 1950. Loess formations of the Mississippi Valley. *Journal of Geology.* 58:599–623.

Leighton, M. M. and Brophy, J. A. 1966. Farmdale glaciation in northern Illinois and southern Wisconsin. *Journal of Geology.* 74:478–499.

Leventer, A. 1981. Relationships between anoxia, glacial meltwater and microfossil preservation-productivity in the Orca Basin, Northwestern Gulf of Mexico. M.S. Thesis, University of South Carolina, Columbia, S.C.

Leventer, A.; Williams, D. F.; and Kennett, J. P. 1982. Dynamics of Laurentide ice sheet during the last glaciation: Evidence from the Gulf of Mexico. *Earth and Planetary Science Letters.* 59:11–17.

Leventer, A.; Williams, D. F.; and Kennett, J. P. 1983. Relationships between anoxia, glacial meltwater and microfossil preservation in the Orca Basin, Gulf of Mexico. *Marine Geology* (in press).

Leverett, F. 1897. The Pleistocene features and deposits of the Chicago area. *Chicago Academy of Science Geology and Natural History Survey Bulletin.* 2:1–86.

Lowrie, W. and Alvarez, W. 1981. One hundred million years of geomagnetic polarity history. *Geology.* 9:392–397.

Mahood, G. A. 1980. Geological evolution of a Pleistocene rhyolitic center—Sierra la Primavera, Jalisco, Mexico. *Journal of Volcanology and Geothermal Research.* 8:199–230.

Malmgren, B. and Kennett, J. P. 1976. Principal component analysis of Quaternary planktonic foraminifera in the Gulf of Mexico: Paleoclimatic applications. *Marine Micropaleontology.* 1:299–306.

Mangerud, J.; Sønstegaard, E.; and Sejrup, H. P. 1979. Correlation of the Eemian (interglacial) Stage and the deep-sea oxygen-isotope stratigraphy. *Nature.* 277:189–192.

Mankinen, E. A. and Dalrymple, G. B. 1979. Revised geomagnetic polarity time scale for the interval 0–5 MYBP. *Journal of Geophysical Research.* 84:615–626.

McCrea, J. M. 1950. On the isotopic chemistry of carbonates and a paleotemperature scale. *Journal of Chemistry and Physics.* 18:849–857.

McDougall, I. 1977. The present status of the geomagnetic polarity time scale. *Pub. No. 1288. Research School of Earth Sciences.* Australian National University. 1–34.

McDougall, I. and Wensink, H. 1966. Paleomagnetism and geochronology of the Pliocene-Pleistocene lavas in Iceland. *Earth and Planetary Science Letters.* 1:322.

McGee, W. J. 1878. On the relative positions of the Forest Bed and associated drift formations in northeastern Iowa. *American Journal of Science.* 15:339–341.

Mellis, O. 1954. Volcanic ash horizons in deep-sea sediments from the eastern Mediterranean. *Deep-Sea Research.* 2:89–92.

Mesolella, K. J.; Matthews, R. K.; Broecker, W. S.; and Thurber, D. L. 1969. The astronomical theory of climatic change: Barbados data. *Journal of Geology.* 77:250.

Milankovitch, M. 1941. Canon of insolation and the Ice-Age problem. *Royal Serbian Academy Special Publication 132.* Section of Mathematical and Natural Sciences. 33. (Translated U.S. Dept. of Commerce.)

Mojski, J. and Ruhle, E. 1965. Geological atlas of Poland stratigraphic and facila problems. *Fascicule 12. Quaternary.* Warsaw.

Moore, W. S. 1982. Late Pleistocene sea-level history. In: *Uranium Series Disequilibrium: Application to Environmental Problems.* Ivanovich, M. and Harmon, R. S., eds. 481–496. Oxford: Clarendon Press.

Morley, J. J.; Hays, J. D.; and Robertson, J. H. 1982. Stratigraphic framework for the late Pleistocene in the northwest Pacific Ocean. *Deep-Sea Research.* 29:1485–1499.

Murray, J. E. 1968. The drift, deterioration and distribution of icebergs in the North Atlantic Ocean. In: *Ice Seminar:* A Conference Sponsored by the Petroleum Society of the Canadian Institute of Mining and Metalurgy (CIM), Calgary, Alberta, Canada. CIM Special Vol. 10, 3–18.

Murray, G. E. 1961. *Geology of the Atlantic and Gulf Coastal Provinces of North America.* New York: Harper Brothers.

Naeser, C. W.; Izett, G. A.; and Wilcox, R. E. 1973. Zircon fisson-tract ages of Pearlette family ash beds in Meade County Kansas. *Geology.* 1:187–189.

Neff, E. 1983. Pleistocene planktonic foraminiferal biostratigraphy and paleoclimatology of the Gulf of Mexico. M.S. Thesis. University of South Carolina. Columbia, S.C.

Ness, G.; Levi, S.; and Couch, R. 1980. Marine magnetic anomaly time scales for the Cenozoic and late Cretaceous: A precise, critique and synthesis. *Reviews of Geophysics and Space Physics.* 18:753–770.

Ninkovich, D. and Heezen, B. C. 1967. Physical and chemical properties of volcanic glass shards from Pozzuolana Ash, Thera Island, and from upper and lower ash layers in eastern Mediterranean deep-sea sediments. *Nature.* 213:582–584.

Ninkovitch, D. and Shackleton, N. J. 1975. Distribution, stratigraphic position, and age of ash layer "L", in the Panama Basin region. *Earth and Planetary Science Letters.* 27:20–34.

Ninkovich, D.; Heezen, B. C.; Conolly, J. R.; and Burckle, L. H. 1964. South Sandwich tephra in deep-sea sediments. *Deep-Sea Research.* 11:605–619.

Ninkovich, D.; Sparks, R. S. J.; and Ledbetter, M. T. 1978. The exceptional magnitude and intensity of the Toba eruption, Sumatra: An example of the use of deep-sea tephra layers as a geological tool. *Bulletin of Volcanology.* 41: 1–13.

Norin, E. 1958. The sediments of the Tyrrhenian Sea. *Reports of the Swedish Deep-Sea Expedition.* 8:4–136.

Obradovich, J. D.; Naeser, C. W.; Izett, G. A.; Pasini, G.; and Bigazzi, G. 1982. Age constraints on the proposed Plio-Pleistocene boundary stratotype at Vrica, Italy. *Nature.* 298:55–59.

Olausson, E. 1965. Evidence of climatic changes in North Atlantic deep-sea cores, with remarks on isotopic paleotemperature analysis. *Progress in Oceanography.* 3:221–252.

Opdyke, N. D. 1972. Paleomagnetism of deep-sea cores. *Review of Geophysics and Space Physics.* 10:213–249.

Opdyke, N. D.; Burckle, L. H.; and Todd, A. 1974. The extension of the magnetic time scale in sediments of the central Pacific Ocean. *Earth and Planetary Science Letters.* 22:300–306.

Opdyke, N. D.; Glass, B.; Hays, J. D. and Foster, J. 1966. Paleomagnetic study of Antarctic deep-sea cores. *Science.* 154:349–357.

Parker, F. L. 1954. Distribution of foraminifera in the northeastern Gulf of Mexico. *Harvard Museum Comparative Zoology Bulletin.* 111:453–588.

——— 1967. Late Tertiary biostratigraphy (planktonic foraminifera) of tropical Indo-Pacific deep-sea cores. *Bulletin of American Paleontology.* 52:115–208.

———— 1973. Late Cenozoic biostratigraphy (planktonic foraminifera) of tropical Atlantic deep-sea sections. *Revista Espanola Micropaleontologia.* 5:253–289.

Pelosio, G.; Raffi, S.; and Rio, D. 1980. The Plio-Pleistocene Boundary Controversy: status in 1979 in light of the International Stratigraphic Guide. *Universita degli studi di Parma. Grafiche Step Editrice.* 131–138.

Phleger, F. B. 1951. Ecology of foraminifera, northwest Gulf of Mexico. *Geological Society of America Memoir 46.* 1:1–88.

Penck, A. and Bruckner, E. 1909. *Die Alpen im Eiszeitalter* (3 vols.). Leipzig: Tauchnitz.

Picard, K. 1964. Die stratigraphische Stellung der Warthe-Eiszeit in Schleswig-Holstein (Deutschland). *Report 6th International Congress of the Quaternary.* 2:191–197.

Pisias, N. G. and Moore, J. C., Jr. 1981. The evolution of Pleistocene climate: A time series approach. *Earth and Planetary Science Letters.* 52:450–458.

Poag, C. W. 1972. Correlation of Early Quaternary events in the U.S. Gulf Coast. *Quaternary Research.* 2:447–469.

Powers, R. B., ed. 1981. *Geologic Frame, Petroleum Potential, Petroleum-Resource Estimates, Mineral and Geothermal Resources, Geologic Hazards, and Deep-Water Drilling Technology of the Maritime Boundary Region in the Gulf of Mexico.* U.S. Geological Survey Open-File Report. 81–265.

Prell, W. L.; Hutson, W. H.; and Williams, D. F. 1979. The subtropical convergence and late Quaternary circulation in the southern Indian Ocean. *Marine Micropaleontology.* 4:225–234.

Prell, W. L. 1980. A continuous high-resolution record of Quaternary evidence for two climatic regions: DSDP hydraulic piston core site 502. *Geological Society of America Abstracts with Programs.* 12:503.

Prell, W. L. and Gardner, J. V. and 12 others 1980. Hydraulic piston coring of late Neogene and Quaternary sections in the Caribbean and equatorial Pacific: preliminary results of Deep Sea Drilling Project Leg 68. *Geological Society of America Bulletin.* 91:433–444.

Prell, W. L. 1982. A re-evaluation of the initiation of Northern Hemisphere glaciation at 3.2 m.y.: New isotopic evidence. *Geological Society of America Abstracts with Program.* 12:592.

———— 1983. Oxygen and carbon isotope stratigraphy for the Quaternary of the Hole 502B: Evidence for two modes of isotopic variability. In: *Initial Reports of the Deep-Sea Drilling Project* vol. 68. Prell, W. L.; Gardner, J. V.; et al., eds. 455–464. Washington, D.C.: U.S. Government Printing Office.

Rabek, K. V. E. 1983. Late Pleistocene Tephrochronology of the Western Gulf of Mexico. M.S. Thesis. University of Georgia. Athens, Ga.

Reed, E. C. and Dreezen, V. H. 1964. Revision of the classification of the Pleistocene deposits of Nebraska. *Nebraska Geological Survey Paper.* 18:1–65.

Reed, E. C.; Dreezen, V. H.; Bayne, C. K.; and Shultz, C. B. 1965. The Pleistocene in Nebraska and northern Kansas. In: *The Quaternary of the United States,* Wright, H. E., Jr. and Frey, D. G., eds. 187–202. International Association of Quaternary Research 7th Congress. Princeton, N.J.: Princeton University Press.

Rezak, R. and Henry, V. J., eds. 1972. *Contributions on the Geological and Geophysical Oceanography of the Gulf of Mexico.* Texas A & M University Oceanographic Studies. vol. 3. Houston: Gulf Publishing Company.

Richardson, D. and Ninkovich, D. 1976. Use of K_2, Rd, Zr, and Y versus SiO_2 in volcanic ash layers of the eastern Mediterranean to trace their source. *Bulletin of the Geological Society of America.* 87:110–116.

Richmond, G. M. 1965. Glaciation of the Rocky Mountains. In: *The Quaternary of the United States.* Wright, H. E., Jr. and Frey, D. G., eds. 217–230. International Association of Quaternary Research 7th Congress. Princeton, N.J.: Princeton University Press.

———— 1970. Comparison of the Quaternary Stratigraphy of the Alps and Rocky Mountains. *Quaternary Research.* 1:3–28.

Rio, D.; Sprovieri, R.; and Raff, I. 1984. *Globorotalia truncatulinoides* in the Mediterranean Pliocene. *Micropaleontology.* In press.

Rona, E. and Emiliani, C. 1969. Absolute dating of Caribbean cores P6304–8 and P6304–9. *Science.* 163:66–68.

Rose, W. I., Jr.; Grant, N. K.; and Easter, J. 1979. Geochemistry of the Los Chocoyos Ash, Quezaltenango Valley, Guatemala. In: *Ash-Flow Tuffs.* Chapin, C. E. and Elston, W. E., eds. 87–89. *Geological Society of America Special Paper* 180.

Rosholt, J. N.; Emiliani, C.; Geiss, J.; Koczy, F. F.; and Wangersky, P. J. 1961. Absolute dating of deep-sea cores by the $^{231}Pa/^{230}Th$ method. *Journal of Geophysical Research.* 67:2907–2911.

Ruddiman, W. F. 1971. Pleistocene sedimentation in the equatorial Atlantic: Stratigraphy and faunal paleoclimatology. *Geological Society of America Bulletin.* 82:283–302.

———— 1977. Late Quaternary deposition of ice-rafted sand in the subpolar North Atlantic (latitude 40° to 65°N). *Geological Society of America Bulletin.* 88:1813–1827.

Ruddiman, W. F. and McIntyre, A. 1976. Northeast Atlantic paleoclimatic changes over the past 600,000 years. In: *Investigation of Late Quaternary Paleoceanography and Paleoclimatology.* Cline, R. M. and Hays, J. D., ed. 111–146. *Geological Society of America Memoir 145.*

———— 1981. Oceanic mechanisms for amplification of the 23,000-year ice volume cycle during the last 250,000 years. *Science.*, 212:617–627.

Ruhe, R. V. and Scholtes, W. H. 1959. Important elements in the classification of the Wisconsin glacial stage: A discussion. *Journal of Geology.* 67:585–593.

Ryan, W. B. F.; Cita, M. B.; Rawson, M. D.; Burckle, L. H.; and Saito, T. 1974. A paleomagnetic assignment of Neogene stage boundaries and the development of isochronous datum planes between the Mediterranean, the Pacific and Indian Oceans in order to investigate the response of the world ocean to the Mediterranean salinity crisis. *Rivista Italiana Paleontologia* 80:631–687.

Saito, T. and Van Donk, J. 1974. Oxygen and carbon isotope measurements of late Cretaceous and early Tertiary foraminifera. *Micropaleontology.* 20:152–177.

Saito, T.; Burckle, L. H.; and Hays, J. D. 1975. Late Miocene to Pleistocene biostratigraphy of equatorial Pacific sediments. In: *Late Neogene Epoch Boundaries.* Saito, T. and Burckle, L. H., eds. 226–244. Micropaleontology Press Special Publication No. 1. New York: American Museum of Natural History.

Salloway, J. C. 1983. Paleomagnetism of sediments from Deep-Sea Drilling Project Leg 71. In: *Initial Report of the Deep Sea Drilling Project.* Ludwig, W. J. and Krasheninnikov, V. A., eds. 71:1073–1091.

Sangree, J. B. 1978. Recognition of continental-slope seismic facies, offshore Texas–Louisiana. *American Association of Petroleum Geologists Studies in Geology.* 7:87–116.

Savin, S. M. 1977. The history of the Earth's surface temperature during the past 100 million years. *Annual Review of Earth and Planetary Sciences.* 5:319–355.

Savin, S. M. and Douglas, R. G. 1973. Stable isotope and magnesium geochemistry of Recent planktonic foraminifera from the South Pacific. *Geological Society of America Bulletin.* 84:2327–2342.

Savin, S. M. and Yeh, H.-W. 1981. Stable isotopes in ocean sediments. In: *The Sea, The Oceanic Lithosphere* vol. 7. Emiliani, C., ed. 1521–1554. New York: John Wiley and Sons.

Savin, S. M.; Douglas, R. G.; and Stehli, F. G. 1975. Tertiary marine paleotemperatures. *Geological Society of America Bulletin.* 86:1499–1510.

Schaefer, I. 1953. Die Donaueiszeitlichen Ablagerungen an Lech und Wertach. *Geologica Bavaria.* 19:13–64.

Schott, W. 1935. Die Foraminiferan in dem aquatorialen Teil des Atlantischen Ozeans: Deutsch. Atlantic Expedition Meteor 1925–1927. *Wissenschaftliche Ergeben* 3:43–134.

Schroder, J. and Dehm, R. 1951. Die Mollusken Fauna aus der Lehmzwischenlage des Deckenschotters von Fischback, Kries Augsburg. *Geologica Bavaria.* 6:118–120.

Self, S. and Sparks, R. S. J. 1981. Tephra Studies. *NATO Advanced Study Series* Vol. 5. Boston: D. Riedel Publishing Company.

Selli, R. 1967. The Pliocene-Pleistocene boundary in Italian marine sections and its relationship to continental stratigraphies. In: *Progress in Oceanography* Vol. 4. Sears, M., ed. 67–86. New York: Pergamon Press.

Selli, R.; Accorsi, C. A.; Bandi, C.; Mazzanti, M.; Bertolani Marchetti, D.; Bigazzi, G.; Bonadonna, F. P.; Boisetti, A. M.; Coti, F.; Colalongo, M. L.; D'Onofrio, S.; Landini, W.; Menisini, E.; Mezzetti, R.; Pasini, G.; Savelli, C.; and Tampieri, R.; 1977. The Urica Section (Calabria, Italy). A potential Neogene/Quaternary boundary stratotype. *Giornale di. Geologia* 42(2):181–204.

Shackleton, N. J. 1969. The past interglacial in the marine and terrestrial records. *Proceedings of the Royal Society London Series B.* 174:135–154.

———— 1977a. The oxygen-isotope stratigraphic record of the Late Pleistocene. *Philosophical Transactions of the Royal Society of London.* 280:169–182.

Shackleton, N. J. and Opdyke, N. D. 1973. Oxygen isotope and paleomagnetic stratigraphy of equatorial Pacific core V28–238: Oxygen isotope temperatures and ice volumes on a 10^5-year and 10^6-year scale. *Quaternary Research.* 3:39–55.

Shackleton, N. J. and Opdyke, N. D. 1976. Oxygen-isotope and paleoemagnetic stratigraphy of Pacific core V28–239 late Pliocene latest Pliocene. In: *Investigation of Late Quaternary Paleoceanography and Paleoclimatology.* Cline, R. M. and Hays, J. D., eds. 449–464. Geological Society of America Memoir 145.

————1977. Oxygen isotope and paleomagnetic evidence for early Northern Hemisphere glaciation. *Nature.* 270:216–219.

Shackleton, N. J. and Matthews, R. K. 1977. Oxygen isotope stratigraphy of late-Pleistocene coral terraces in Barbados. *Nature.* 268:618–620.

Shackleton, N. J. and Cita, M. B. 1979. Oxygen and carbon isotope stratigraphy of benthic foraminifers at Site 397: detailed history of climatic change during the Neogene: In: *Initial Reports of the Deep Sea Drilling Project.* vol. 47. von Rod, U.; Ryan, W. B. F.; et al., eds. 743–755. Washington, D.C.: U.S. Government Printing Office.

Shackleton, N. J.; Backman, J.; Kent, D.; Imbrie, J.; Pestiaux, P.; Pisias, N. G.; and Zimmerman, H. B. 1982. The evolution of climate response to orbital forcing: results over three million years from DSDP Site 552A. *International Symposium on Milankovitch and Climate Program and Abstracts.* New York: Palisades.

Shackleton, N. J. and Hall, M. A. 1983. Stable isotope record of hole 504 Sediments: high resolution record of the Pleistocene. In: *Initial Reports of the Deep Sea Drilling Project* v. 69. Cann, J. R.; Langseth, M. G.; et al., eds. 431–441. Washington, D. C.: U.S. Government Printing Office.

Shilts, W. W.; Miller, G. H.; and Andrews, J. T. 1981. Glacial flow indicators and Wisconsin glacial chronology, Hudson Bay/James Bay Lowlands: Evidence against a Hudson Bay ice divide. *Geological Society of America Abstracts with Programs.* 13:553.

Shimek, B. 1909. Altonian sands and gravels in western Iowa. *Geological Society of America Bulletin.* 20:399–408.

Shokes, R. F.; Trabant, P. K.; Presley, B. J.; and Reid, D. F. 1977. Anoxic hypersaline basin in the northern Gulf of Mexico. *Science.* 196:1443–1446.

Snyder, S. W. 1978. Distribution of planktonic foraminifera in surface sediments of the Gulf of Mexico. *Tulane Studies in Geology and Paleontology* 14:1–80.

Stromer, E. and Lebling, C. 1929. Fossilführendes Pliozän in Südbayern. *Zentralblatt für Mineralogie Jahrgang 1929.* Abteilung B:307–314.

Tauxe, L.; Opdyke, N. D.; Pasini, G.; and Elmi, C. 1983. Age of the Plio-Pleistocene boundary in the Vrica Section, southern Italy. *Nature.* 304:125–129.

Theyer, F. and Hammond, S. 1974. Paleomagnetic polarity sequence and radiolarian zones, Brunhes to Epoch 20. *Earth and Planetary Science Letters.* 22:306–319.

Thierstein, H. R.; Geitzenauer, K. R.; Molfino, B.; and Shackleton, N. J. 1977. Global synchroneity of late Quaternary coccolith datum levels: validation by oxygen isotopes. *Geology.* 5:400–404.

Thompson, P. R. and Sciarrillo, J. R. 1978. Planktonic foraminiferal biostratigraphy in the Equatorial Pacific. *Nature.* 275:29–33.

Thompson, P. R.; Bé, A. W. H.; Duplessy, J. C.; and Shackleton, N. J. 1979. Disappearance of pink-pigmented *Globigerinoides ruber* at 120,000 yr B.P. in the Indian and Pacific Oceans. *Nature.* 280:554–558.

Thunell, R. C. 1976. Calcium carbonate dissolution history in late Quaternary deep-sea sediments, western Gulf of Mexico. *Quaternary Research.* 6:281–297.

——— 1981a. Cenozoic paleotemperature changes and planktonic foraminiferal speciation. *Nature.* 289:670–672.

——— 1981b. Late Miocene-early Pliocene planktonic foraminiferal biostratigraphy and paleoceanography of low latitude marine sequences. *Marine Micropaleontology.* 6:71–90.

Thunell, R. C. and Williams, D. F. 1983. The stepwise development of Pliocene-Pleistocene paleoclimatic and paleoceanographic conditions in the Mediterranean. *Utrecht Micropaleontological Bulletin.* In press.

Thunell, R. C.; Williams, D. F.; and Kennett, J. P. 1977. Late Quaternary paleoclimatology, stratigraphy and sapropel history in eastern Mediterranean deep-sea sediments. *Marine Micropaleontology.* 2:371–388.

Thunell, R. C.; Federman, A.; Sparks, S.; and Williams, D. F. 1979. The age, origin, and volcanological significance of the Y-5 ash layer in Mediterranean. *Quaternary Research.* 12:241–253.

Tompkins, R. E. and Shephard, L. E. 1979. Orca Basin: Depositional processes, geochemical properties and clay mineralogy of Holocene sediments within an anoxic, hypersaline basin, northwest Gulf of Mexico. *Marine Geology.* 33:221–238.

Trabant, P. K. and Presley, B. J. 1978. Orca Basin: An anoxic depression on the continental slope, northwest Gulf of Mexico. In: *Framework, Facies, and Oil-Trapping Characteristics of the Upper Continental Margin.* Bouma, A. H.; Moore, G. T.; and Coleman, J. M., eds. 289–303. American Association of Petroleum Geologists Studies in Geology. no. 7. Tulsa, Okla.: American Association of Petroleum Geologists.

Turner, C. 1975. The correlation and duration of Middle Pleistocene interglacial periods in Northwest Europe. In: *After the Australopithecines*. Butzer, K. W. and Isaac, G. L., eds. 259–308. The Hague: Mouton.

Uchupi, E. and Emery, K. O. 1968. Structure of continental margin off Gulf Coast of United States. *American Association of Petroleum Geologists*. 52:1162–1193.

Urey, H. C. 1947. The thermodynamic properties of isotopic substances. *Journal of the Chemical Society*. Part I 562–581.

Urey, H. C.; Lowenstam, H. A.; Epstein, S.; and McKinney, C. R. 1951. Measurement of paleotemperatures and temperatures of the Upper Cretaceous of England, Denmark, and the Southeastern United States. *Geological Society of America Bulletin*. 62:399–416.

Vallier, T. L. and Kidd, R. B. 1977. Volcanogenic sediments in the Indian Ocean. In: *Indian Ocean Geology and Biostratigraphy. Studies Following Deep-Sea Drilling Legs 22–29*. Heirtzler, J. R. et al., ed. Washington, D.C.: American Geophysical Union.

Van der Hammen, T.; Wijmstra, T. A.; Zagwijn, W. H. 1971. The floral record of the late Cenozoic of Europe. In: *Late Cenozoic Glacial Ages*. Turekian, K. K., ed. 391–424. New Haven: Yale University Press.

Van Donk, J. 1976. ^{18}O record of the Atlantic Ocean for the entire Pleistocene Epoch. In: *Investigation of Late Quaternary paleoceanography and paleoclimatology*. Cline, R. M. and Hays, J. D., eds. 147–163. *Geological Society of America Memoir 145*.

Van Hinte, J. E. 1976a. Jurassic time scale. *American Association of Petroleum Geologists Bulletin*. 60:489–497.

———— 1976b. A Cretaceous time scale. *American Association of Petroleum Geologists Bulletin*. 60:498–516.

Veeh, H. H. and Chappell, J. M. A. 1970. Astronomic theory of climatic change: Support from New Guinea. *Science*. 167:862.

Vergnaud-Grazzini, C.; Ryan, W.; and Cita, M. B. 1977. Stable isotopic fractionation of carbon and oxygen in benthic foraminifera. *Earth and Planetary Science Letters*. 8:247–252.

Vergnaud-Grazzini, C.; Grably, M.; Pujol, C.; and Duprat, J. In press. Oxygen isotope stratigraphy and palaeoclimatology of Southwest Atlantic Quaternary Sediments (Rio Grande Rise) at DSDP Site 517. In: *Initial Reports of the Deep Sea Drilling Project*. Vol. 72. Barker, P. F.; Carlson, R. L.; et al., eds. Washington, D.C.: U.S. Government Printing Office.

Vincent, E. and Berger, W. H. 1981. Planktonic foraminifera and their use in paleoceanography. In: *The Sea*. Vol. 7. Emiliani, C., ed. New York: Wiley and Sons.

Vincent, E.; Killingley, J. S.; and Berger, W. H. 1961. Stable isotopes in benthic foraminifera from Ontong-Java Plateau, box cores ERDC 112 and 123. *Palaeogeography, Palaeoclimatology and Palaeoecology*. 33:221–230.

Vine, F. J. 1966. Spreading of the ocean floor: New evidence. *Science*. 154:1405–1415.

Von Huene, R.; Cronch, J.; and Larson, E. 1976. Glacial advances in the Gulf of Alaska area implied by ice-rafted material. In: *Investigation of Late Quaternary Paleoceanography and paleoclimatology*. Cline, R. M. and Hays, J. D., eds. 411–412. *Geological Society of America Memoir 145*.

Wahrhaftig, C. and Birman, J. H. 1965. The Quaternary of the Pacific Mountain system in California. In: *The Quaternary of the United States*. Wright, H. B., Jr. and Frey, D. G., eds. 299–339. International Association of the Quaternary Research 7th Congress. Princeton: Princeton University Press.

Warnke, D. A. 1982. Distribution of glacial-marine sediments and the beginning of ice-rafting in the Norwegian-Greenland seas. *American Quaternary Association Program and Abstracts.* 175.

Watkins, N. D. 1972. Review of the development of the geomagnetic polarity time scale and discussion of prospects for its finer definition. *Geological Society of America Bulletin.* 83:551–574.

Watkins, N. D.; Kester, D. R.; and Kennett, J. P. 1974. Paleomagnetism of the type Pliocene/Pleistocene boundary section at Santa Maria de Catanzaro, Italy, and the problem of postdepositional precipitation of magnetic minerals. *Earth and Planetary Science Letters.* 24:113–119.

Watkins, N. D.; Sparks, R. S. J.; Sigurdsson, H.; Huang, T. C.; Federman, A.; Carey, S.; and Ninkovich, D. 1978. Volume and extent of the Minoan tephra from Santorini Volcano: New evidence from deep-sea sediment cores. *Nature.* 271:122–126.

Williams, D. F. and Thunell, R. C. 1979. Faunal and oxygen isotopic evidence for surface water salinity changes during sapropel formation in the eastern Mediterranean. *Sedimentary Geology.* 23:81–93.

Williams, D. F. and Healy-Williams, N. 1980. Oxygen isotopic-hydrographic relationships among Recent planktonic foraminifera from the Indian Ocean. *Nature.* 283:848–852.

Williams, D. F.; Sommer, M. A.; and Bender, M. L. 1977. Carbon isotopic compositions of Recent planktonic foraminifera of the Indian Ocean. *Earth and Planetary Science Letters.* 36:391–403.

Williams, D. F.; Moore, W. S.; and Fillon, R. H. 1981. Role of glacial Arctic Ocean ice sheets in Pleistocene oxygen isotope and sea level records. *Earth and Planetary Science Letters.* 56:157–166.

Willman, H. B. and Frye, J. C. 1970. Pleistocene Stratigraphy of Illinois. *Illinois Geological Survey Bulletin.* 94:204.

Wilson, L.; Sparks, R. S. J.; Huang, T. C.; and Watkins, N. D. 1978. The control of volcanic column heights by eruption energetics and dynamics. *Journal of Geophysical Research.* 83:1829–1836.

Woillard, G. M. and Mook, W. G. 1982. Carbon-14 dates at Grand Pile: Correlation of land and sea chronologies. *Science.* 259:159–161.

Worzel, J. L. 1959. Extensive deep-sea sub bottom reflections identified as white ash. *U.S. National Academy of Science Proceedings.* 45:349–355.

Wright, J. V. 1981. The Rio Caliente Ignimbrite: Analysis of a compound intraplinian ignimbrite from a major late Quaternary Mexican eruption, *Bulletin of Volcanology.* 44:189–212.

Zagwijn, W. H. 1960. Aspects of the Pliocene and Early Pleistocene vegetation in the Netherlands. *Mededel. Geol. Sticht.,* Ser. C-3, 5:1–5.

———— 1975. A model-theory for the Pliocene/Pleistocene boundary determination, based on past climatic changes. In: *Late Neogene Epoch Boundaries.* Saito, T. and Burckle, L. H., eds. 71–74. Micropaleontology Press Special Publication No. 1. New York: American Museum of Natural History.

Zeuner, F. W. 1945. *The Pleistocene Period, Its Climate, Chronology, and Faunal Successions.* London: The Ray Society.

Index